ROBOT,

TAKE THE WHEEL

ROBOT,

TAKE THE WHEEL

THE ROAD TO AUTONOMOUS CARS
AND THE LOST ART OF DRIVING

JASON TORCHINSKY
FOREWORD BY BEAU BOECKMANN

APOLLO
PUBLISHERS

Apollo Publishers books may be purchased for educational, business, or sales promotional use. Special editions may be made available upon request. For details, contact Apollo Publishers at info@apollopublishers.com.

Visit our website at www.apollopublishers.com.

Library of Congress Cataloging-in-Publication Data is available on file.

Print ISBN: 978-1-948062-97-8
Ebook ISBN: 978-1-948062-27-5

Printed in the United States of America.
First printed in hardcover in 2018.

Cover and interior design by Jason Torchinsky and Rain Saukas.
Chapter opener illustrations by Jason Torchinsky.

For Sally, because I love her more than anything. Even all the cars. And Otto, because he's a little kook and I love him too.

Also, thanks to everyone at Jalopnik for being great.

CONTENTS

FOREWORD by Beau Boeckmann,
President and Chief Operating Officer of Galpin Motors 9

INTRODUCTION 15

CHAPTER 1: We've Been Here Before 21

CHAPTER 2: How Did We Get Here? 35

CHAPTER 3: How Do They Work, Anyway? 73

CHAPTER 4: Semiautonomy is Stupid 93

CHAPTER 5: They're Robots, Not Cars 107

CHAPTER 6: Ethics, Behavior, and Being Better than People Are . . . 127

CHAPTER 7: They Shouldn't Look like Cars 151

CHAPTER 8: The Death of the Journey 173

CHAPTER 9: Will They Be Like Your Dog? 189

CHAPTER 10: Save the Gearheads 217

NOTES 237

BEAU BOECKMANN, PRESIDENT AND CHIEF OPERATING OFFICER OF GALPIN MOTORS

WHY WOULD AN AUTOMOBILE DEALER BE ASKED TO WRITE A foreword for a book about autonomous vehicles, which some people say will cause an automobile industry apocalypse? Great question. I think it has something to do with the fact that Jason is a little nuts and we have a lot in common, like our love of unusual cars and obscure automotive history, and our interest in discussing where the automotive industry is heading. While many car guys and gals fear that "the end is near" for driving and car culture—believing that robot taxis will soon take over our roads—the truth is that there is an incredibly exciting future in the autonomous car world that awaits all of us—automotive enthusiasts and haters alike.

A little background about myself: I grew up in Los Angeles and my roots here are deep. My great-grandfather moved to LA in 1879, so you can say I'm a native. LA is an interesting place. Someone once said that God took heaven and hell, mixed them together, and called it Los Angeles. The city definitely has elements of both, especially for a driver. There is nothing more intoxicating than driving a convertible with the top down on the Pacific Coast Highway, or more thrilling than driving a sports car through the twisting roads of Mulholland Highway. But there's also LA traffic—soul-crushing, mind-numbing, blood pressure-spiking traffic. In LA we don't measure by distance (I had no idea how far places were before using the navigation app Waze), but by how much time it takes to get somewhere, which is heavily influenced by the time of day you're driving there.

As far as my work background, I am probably one of the few people who went to college to become a car salesperson. I was lucky enough to grow up with the ultimate automotive-influenced upbringing, at least in the car dealer sense. My father started at Galpin Ford in 1953 as a salesperson. He got promoted and grew Galpin Ford to be the number one volume Ford dealership in the world. He achieved this by caring for customers and employees, working hard, being honest, and being creative, looking for fun, exciting ways to exceed customers' expectations. One of the ways he did this was customizing—or as we call it, "Galpinizing"—vehicles. In 1952, we built our first "Galpin Custom" from a brand new convertible at the dealership. It was shown at the Autorama, named a Top 10 Custom of the Year, and was featured on the cover of *Motor Trend* magazine in June 1953.

While Galpin has done all kinds of aftermarket customizing, racing, performance, off-roading, and more, one thing that gained us particular fame was helping to pioneer and launch the conversion van. In fact, many people credit Galpin for starting the conversion van industry. With conversion vans we weren't just building regular vans (it was the 1970s after all), but ones with wild interiors and themes like Madam Frenchy's—a provocative design complete with striking red fleur-de-lis wallpaper, as well as a love seat, chandelier, and fireplace. The conversion vans were crazy and fun, and their sales went through the roof—with the addition of the conversion vans our van sales went up 500 percent.

It was an interesting time for me. At the shop, I got to hang out with the guys who were designing the wild vehicles. My mother, Jane, an interior designer and businessperson who worked for increasingly sophisticated buyers, even helped with the designs. But the conversion vans proved to be a passing trend, and by the 2000s they were all but a memory.

The first car I customized was one my grandmother willed me—her old 1965 Mercedes 220SE (the four-door kind, not the cool one)—and I loved it. Later, I had the joy of launching Galpin's customizing division, Galpin Auto Sports, and was then invited to join Ford's Product Committee, which looks behind the scenes of product development. This committee has strong input on upcoming products in the near and long term. Many times I witnessed a car go from a sketch to full production. It was something to behold.

As a member of the Product Committee I heard about autonomous vehicles pretty early, and since then I've had numerous conversations on the future of autonomous vehicles and how they will impact society. This is not just related to customers and drivers, but the entire automotive industry and business world as well. Could automated vehicles be the next great threat, the one that finally knocks car dealers out of business? Many people think car manufacturers won't survive the automated car revolution, and when I first began considering this, I had to do some real soul-searching. Should I walk away from the business and my passion, or should I double down against an industry change, from people driving cars they're passionate about to everyone riding around in soulless robo-taxis?

The truth is that I don't believe it has to be one or the other. As an enthusiast and someone whose career is on the line, I can honestly say that I am excited about our autonomous automotive future.

As consumers, autonomous vehicles will bring us choices. Taxis and public transportation will be revolutionized. The car business and dealers are going to need to adapt. I'm geeked to focus back on what could prove to be the perfect automated vehicle: the van. I can't wait to get back in the conversion van business—this time using them as robo-taxis. Some vehicles will have both full (human-controlled) driving and autonomous modes. You could spend a whole day driving your heart out, and then let the car take you home in rush hour traffic while you relax and catch up on things. That's taking the heaven and hell out of LA driving and making the best out of both.

What I know about the world of automated cars, after all my studying and "peek behind the curtain," leads me to know that Jason is the best predictor yet on the uncertain autonomous future. He's informed about the car industry and points out what he doesn't know (unlike so many industry "experts" and newcomers, who make claims without factual support), and considers fascinating new opportunities that autonomous vehicles could bring in the not-so-distant future.

The introduction of autonomous vehicles is not going to be easy, and it's going to be uncomfortable for a while, (and don't worry, Mom, I'm not going to force you to ride with me and a robot driver). But it's probably going to be quite a while before we see any real impact on society. There's no need to worry because, while technology is going to change things in some very dramatic ways, cars with human drivers are here to stay. People love cars, and they love driving, and that isn't going to change. When people discuss automated vehicles they forget our irrational love for the automobile, and how important it is to our culture. And while the automotive business is not always the easiest, it's one that perseveres, and if you love people (and at least moderately like cars) it's a wonderful business to be in. For me, working at events like car shows can blur the line between work and play. Even when automated cars become part of the business, I'm not going anywhere.

This book is not about an enthusiast trying to justify the future, or an industry or media that wants to believe—depending on which side you're on—in an automotive dystopia or utopia. That's why this book is such a fun ride; it's an honest look at what may (or may not) lie ahead in our autonomous car future. Either way, it's going to be one hell of a ride—or drive!

INTRODUCTION

EMPTY YOUR POCKETS.

I'm going to bet that among the wadded-up receipts, an effectively valueless amount of change, and something that may have once been gum, you'll also find a small, powerful, handheld computer. Let's think about what you call that computer. In the US, you probably call it a "phone," and in most of the rest of the world you likely call it a "mobile," a truncated form of "mobile phone."

Even though the actual business of voice-based telephoning is just one of the millions of things you could be doing on your device and is likely not even the most common thing you use your device for, the name has stuck. We call these machines "phones" because when they first started to become something that normal, non-jet-owning people could own or use in the early 1990s, that's really all they were. They were portable phones.

If, in the early 1990s, you were paying attention to these portable phones, and you were the bright, thoughtful person you are today, then I bet you could have easily imagined a future where everyone had their own personal cell phone, ready to take calls anytime, anywhere. The world you could have imagined would have been a big improvement over the real world, plush as it was with those miserable, wall-tethered boat anchors we used to make calls on. A world where we each have our own personal, portable phone would have been a smart, reasonable extrapolation of the world as you knew it.

Of course, as we know now, you would have been totally wrong. What portable phones became is not something most people could have predicted. Very few people looked at the crude, brick-like portable phones of the early 1990s, with their one-line numeric displays, and imagined that, someday, these devices would become the primary terminals for people to access the small but growing network of government and university computers known as the Advanced Research Projects Agency Network (ARPANET), and that through this network people would use their pocket-computer terminals to read magazines, send short messages and longer letters to one another, use integrated cameras to take photographs, broadcast video to a global audience, read short, strange missives from the president of the United States and comment back on them, consume television shows, pornography, and movies, and send pictures of their own genitals to people, possibly destroying their careers in the process.

Nobody in the early 1990s imagined that this would be where those clunky portable phones would take us, and yet here we are.

When it comes to autonomous cars, for most of us, people and companies, it's 1990 all over again. This time, instead of portable phones, we're talking about autonomous cars—but we're still imagining the future the same way: like now, but better. If portable phones have taught us anything, it's that we're really bad at predicting where new technologies will lead us.

Most automakers developing autonomous vehicles, which is pretty much every major car manufacturer, is still thinking of what they're building as cars. That's because, at the moment, that's exactly what they are: cars that are learning to drive themselves. All semiautonomous cars being sold today, from Tesla or Volvo or Mercedes-Benz or whomever, are based on cars originally designed for human drivers, augmented with sensors and computers to allow for some, quite limited, degree of driving autonomy. Right now, we're at the bag phone stage (remember those? We don't seem to put new tech in bags anymore); or at best, the brick phone stage. These machines are effectively doing the same job as their predecessors, but have one key new trait: for phones it was portability, for cars it's self-driving.

If we want to get a sense of what the future may hold, and how that future may affect us and our culture, we need to start looking at autonomous cars as something separate from cars. If we take a step back to get a wider perspective, we can see that once fully autonomous cars are developed and sold to the public in a meaningful quantity, this will represent the first truly large deployment of large-scale, highly mobile robots into human society. These are not Roombas—scuttling about under couches, foraging for Dorito fragments—but machines weighing close to two tons, fully capable of ending a human life.

I'm not trying to be an alarmist here; cars have been capable of ending human lives for well over a century, but until now only at the hands of human pilots. Besides, autonomous cars will probably save more lives than they'll take; one of the effects of their wide-scale deployment will likely be less loss of human life, because the cars will drive better and more safely than we do. But just as today's cell phones are so much more than just phones, autonomous cars are going to be so much more than just cars, and we may as well accept that now.

This book is about the coming age of autonomous cars and is an attempt to get you to consider them as something beyond cars as we understand them today. It's not a book about the details of the technology, because that changes so fast and so many people so much smarter than me can write those books. This book is essentially a giant thought experiment, where we'll try and imagine what the coming of autonomous vehicles means to us; how we'll get along with the robots that will take over our cars' jobs; what these things will look like; what sorts of jobs they may do; what we can expect of them; how they should act, ethically; how we can have fun with them; and how those of us who love to drive, manually and laboriously, can continue to do so.

It's probably worth pointing out just what sort of a book this will be. If you're looking for something crammed full of the latest facts, statistics, and research about autonomous cars and their development, and up-to-the-minute information about the current state-of-the-art cars, this isn't that book. If you want that, look on the internet. It gets updated far more often than books do, and you'll be much happier. I don't want to compete with the internet for anything like that, because I'll lose.

This book also doesn't reach out to many experts, despite how often PR people and agents for these experts like to email me. I'm not ignoring the experts in the field out of any disrespect, but the truth is that the full impact of autonomous cars isn't even close to being felt. Even if an expert has more degrees than a thermometer, and despite however closely they're working with this or that autonomous car start-up with acres of venture capital funding, they're going to be pulling guesses *ex recto*, just like I am. So I'm just going to give it a go myself, because why not?

Think about this book like that—some guy, we'll call him "me," is interested in cars and robots and the culture surrounding both, and is thinking a lot about it and asking a lot of questions, not all of which he has answers to or can even pretend he has answers to.

Because I don't. But the questions are still worth asking, and it's still worth thinking about how things could be, how we want them to be, and how we're afraid they may end up. This is a conversation about what autonomous cars may be or mean or become, and if you're reading this at some point in the future, laughing about how wrong I was about *everything*, I can't say I'll be too shocked.

This is an exciting era we're in. Autonomy will be the biggest shift in how we interact with our cars in decades, and it's going to reshape how we transport ourselves more than any other advancement in recent memory. It's going to end up far, far weirder than we think, I'm pretty sure, so we may as well get a head start and think some things through.

Don't worry. It'll be fun.

CHAPTER 1

WE'VE BEEN HERE BEFORE

FOR ALL THE EXCITEMENT AND HYPE SURROUNDING AUTONO-
mous vehicles, it's worth remembering that, for most of the history of mankind, we've been using vehicles that were capable of full autonomy. We call these vehicles "horses" or sometimes "donkeys" or "camels" or any number of other large, muscular mammals that we've coerced into taking us from place to place. All of these are, of course, fully autonomous, and have been for thousands and thousands of years before any horse ever even saw a human.

Generally, horses and other animals squander their autonomy wandering around, eating lawns, having steamy horse sex and making new horses to start the whole thing over again. Once employed by humans for the purpose of transport, animals like horses became, effectively, semiautonomous vehicles.

There's actually an accepted system in place for describing levels of autonomy for cars, known as the SAE (that's Society of Automotive Engineers, like the Freemasons but much worse dressers) automation levels. They break down like this:

Level 0: No automation, the human driver does all the driving.

Level 1: Driver assistance, an advanced driver assistance system (ADAS) on the vehicle can sometimes assist the human driver with either steering or braking/accelerating, but not both simultaneously.

Level 2: Partial automation, an ADAS on the vehicle can actually control both steering and braking/accelerating simultaneously under some circumstances. The human driver must continue to pay full attention ("monitor the driving environment") at all times and perform the rest of the driving task.

Level 3: Conditional automation, an automated driving system (ADS) on the vehicle can perform all aspects of the driving task under some circumstances. In those circumstances, the human driver must be ready to take back control at any time when the ADS requests the human driver to do so. In all other circumstances, the human driver performs the driving task.

Level 4: High automation, an ADS on the vehicle can perform all driving tasks and monitor the driving environment— essentially, do all the driving—in certain circumstances. The human driver need not pay attention in those circumstances.

Level 5: Full automation, an automated driving system on the vehicle can do all the driving in all circumstances. The human occupants are just passengers and need never be involved in driving.

Based on our modern scales, I'd have to say a vehicle composed of a horse and cart is somewhere between Levels 3 and 4: the "vehicle" is in complete control, but some human intervention is required.

Of course, the manner in which a horse is autonomous is quite different from an electromechanical car. While the destination is

SAE AUTONOMY LEVELS

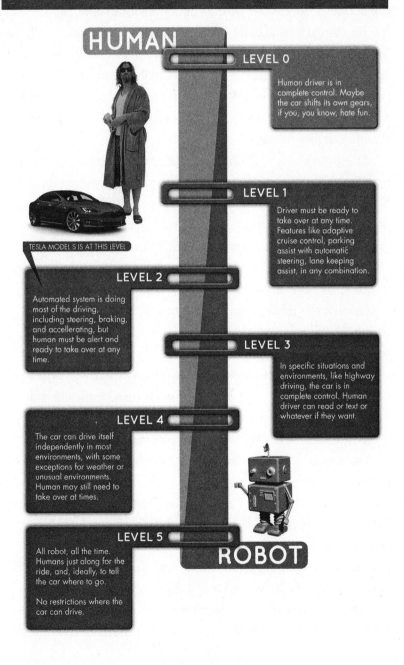

HUMAN

LEVEL 0
Human driver is in complete control. Maybe the car shifts its own gears, if you, you know, hate fun.

LEVEL 1
Driver must be ready to take over at any time. Features like adaptive cruise control, parking assist with automatic steering, lane keeping assist, in any combination.

TESLA MODEL S IS AT THIS LEVEL

LEVEL 2
Automated system is doing most of the driving, including steering, braking, and accellerating, but human must be alert and ready to take over at any time.

LEVEL 3
In specific situations and environments, like highway driving, the car is in complete control. Human driver can read or text or whatever if they want.

LEVEL 4
The car can drive itself independently in most environments, with some exceptions for weather or unusual environments. Human may still need to take over at times.

LEVEL 5
All robot, all the time. Humans just along for the ride, and, ideally, to tell the car where to go.

No restrictions where the car can drive.

ROBOT

pretty much a given for an autonomous car, thanks to GPS, the horse doesn't necessarily know it. What a horse does know are the fundamental mechanics of driving. A horse inherently knows how to stay on a road, follow a path, avoid obstacles, stop if confronted with confusion or danger, make turns, look for potential hazards, and so on. What the horse relies on the human for are inputs regarding the desired speed of travel and guidance to maintain a proper path.

With a horse-car, you're not "steering" the horse in the same way that you steer a car—the horse is handling those mechanics. You're guiding the animal to your destination, and, in some cases, the horse may even know familiar routes and paths, so what the driver needs to do in a horse-car can be pretty minimal.

We forget just how much natural processing an equine brain is doing to drag a streetcar along a path—it's essentially what we're currently trying to make automated vehicles (AVs) do. It should remind us that getting a car safely to your desired destination requires a very specific set of skills, and there's no reason to assume that, as humans, we're somehow hardwired to know how to do it. In fact, the fates of the two earliest human-driven automobiles speak directly to how unprepared we were, and how difficult the basic task of piloting a moving machine really is.

An automobile is any self-propelled wheeled machine designed to transport passengers and/or cargo. What powers that car—as long as it's mechanical in nature, somehow, doesn't really matter. A steam car is as much an automobile as a gasoline, diesel, or electric car. I want to make that abundantly clear in case anyone reading over your shoulder decides to pedantically correct this book. If someone does, please tell them to get bent.

The first machine that we can really call an automobile—a self-propelled, mechanical, wheeled machine driven by a human—was Nicolas-Joseph Cugnot's 1769 steam dray.

(I know Mercedes-Benz likes to talk about how they invented the car; they cite the 1886 Benz Patent-Motorwagen as the first example. Don't be taken in by this self-serving bit of historical revisionism.)

Cugnot's steam dray was designed to be an artillery-hauling truck, basically, and as such was designed in a way that made handling pretty terrible. Really, you probably couldn't design a worse vehicle to drive than Cugnot did, but, in his defense, no one had any idea what the hell "handling" was or what "driving" would be like, or anything at all like that. These problems simply didn't exist before Nick-Joe fired up the huge, teapot-like boiler on the steam dray.

This first car, being designed to haul heavy artillery, cannonballs, and other massive iron things, was designed with all the mechanical parts (and weight) well up front, with a large, flatbed-like area at the rear. The solitary and massive front wheel was driven by steam pistons, and in front of the wheel hung the massive, heavy boiler.

The driver of this thing was expected to steer with a set of handlebars that looked like a steampunk bull's horns, and that driver would likely need the strength of a steampunk bull to be effective. The vehicle was designed to be balanced when piled high with cannonballs or towing cannon. Laden, the balance would likely have been better, but the whole thing would have been so heavy as to be deeply ungainly. Unladen, it would have been lighter, but with all the weight on the one front wheel, steering would have been a nightmare.

Cugnot not only invented the automobile, he invented lethal understeer.

Understeer, when a car turns less sharply than desired, is what happens with nose-heavy, front-wheel drive cars because they naturally want to go in straight lines. Cugnot's steam dray was a ridiculous caricature of this design, and as a result, the first test ended up with Cugnot driving it into a wall, which he partially demolished. The second test didn't fare much better; the truth is that I doubt the steam dray could have been driven effectively. The design was far too unforgiving and difficult and, what's more, nobody had any idea how to drive.

The next attempt at an automobile was built by William Murdoch in 1784 and seemed to recognize the layout issues that Cugnot's vehicle had, and pretty much corrected them. Too bad it only existed as a subscale working model. If it had been built to human scale, it's likely it would have been far more drivable than the Cugnot car.

In 1801, the invent-cars project was renewed with the help of a Cornish man named Richard Trevithick who built a crude but full-scale test vehicle, the Puffing Devil. In 1803, he built a much more realized vehicle, arguably the very first passenger car designed to be a passenger automobile from the start, the London Steam Carriage.

The Puffing Devil was a proof-of-concept test of locomotion; it didn't really have any steering mechanism, or a real passenger compartment. It wasn't "driven" in the sense we understand driving today, which is why we should focus on the London Steam Carriage, which had an actual steering mechanism and a place for passengers. It was a real car, and as such could be driven. Its steam engine was set low in the tall chassis and toward the rear, controlling the rear wheels. A lone steering wheel up front was turned via a simple tiller. It was a basic design, but it was effective. The center of gravity was pretty low for such a tall vehicle and the steering mechanism worked, even if it required two people—one to stoke and manage the engine at the rear, and one to steer up front.

This division of labor necessitated communication between the two parties—as if you, while driving, had to yell at your feet to get off the gas pedal and get on the brake. Even with the task of driving divided between two people—who didn't know how weird that would one day seem because nobody had ever done this before—the act of driving proved difficult.

To Trevithick and his team's credit, they did manage to drive it a bit on the first try, about 10 miles, at speeds between 4 and 9 mph, but the next night they managed to wreck it.

There is a pretty good recounting of the wreck in the *Life of Richard Trevithick: With an Account of His Inventions, Volume 1*.[1]

> They kept going on for four or five miles, and sometimes at
> the rate of eight or nine miles an hour. I was steering, and
> Captain Trevithick and some one else were attending to the
> engine. . . . She was going along five or six miles an hour,
> and Captain Dick called out, "Put the helm down, John!"
> and before I could tell what was up, Captain Dick's foot was
> upon the steering-wheel handle, and we were tearing down
> six or seven yards of railing from a garden wall. A person put
> his head from a window, and called out, "What the devil are
> you doing there! What the devil is that thing!"

What we see here is that people were starting to learn just how much attention and processing is involved in driving, something horses have understood for centuries. A horse pulling a carriage the same size as this 1803 car would not have made this mistake. From what the accident reports state, it looks like the driver misjudged the distance to the garden fences and sideswiped them.

It's a pretty rookie mistake, but, to be fair, the driver of that steam carriage had more driving experience than anyone else on earth.

I mean, if you really think about what was being asked of these early, early drivers, the demands were decidedly nontrivial. For the first time, a vehicle moving at speeds significantly faster than a walking pace had to be controlled through city streets. This means that people had to make many new and unexpected decisions at a pace greater than they'd been used to.

The vehicle itself was tall and ungainly, with extremely skinny, metal-clad wheels that likely had very poor grip. Wheels like these over cobblestone or macadamized streets wouldn't be easy for a modern driver with decades of experience, let alone people new to the fundamental concept of motorized motion. Everything must have felt unfamiliar and strange; the single-wheel steering couldn't have been that confidence inspiring, and understanding how to follow the track of a road is the sort of thing that's only really learned by visceral, physical experience. It comes quickly, but it's not necessarily instant, and in a vehicle as ungainly as the London Steam Carriage, there's a lot about how the car feels and behaves on the road that has to be learned.

All of this is to say that I'm not the least bit surprised that the first automobile drive of any length ended up in a wreck.

Keep in mind that these early cars even predate trains, which diverged from the automobile evolutionary line the year after the London Steam Carriage, with Trevithick's rail locomotive of 1804. The fact that Trevithick, who was part of the second automobile wreck in human history, decided to eliminate most of the driving skill required by running his automobile on rails, is telling. Driving, even though it has become second nature to most of us, isn't easy.

Railroads are a form of mechanical driving semi-automation;

the rails take over the steering, navigation, and lane-keeping duties of a vehicle, a significant portion of the driving task. We went from the semiautonomy of animal power to a brief flirtation with entirely manual driving, then quickly retreated to a new, mechanized form of partial autonomous travel.

Sure, there were plenty of other reasons why railroads became the first widely used system of mechanized travel—poor road networks, economies of scale, centralized ownership, and so on—but the fact that no one knew how to drive is an under-appreciated factor.

I know it feels like we're on the fringes of a revolution in driving, where we're finally free to relinquish control over to a competent, well-trained machine, but the truth is that we're really just going back to where we've been throughout most of history, just in a much more technologically refined way. It is full human control of a body- and ability-enhancing prosthetic—a car—that's the really fascinating development, and it's possible that this past century or so of widespread driving may be the anomaly.

This idea that a human-driven car is essentially a body- and ability-enhancing prosthetic is a concept that's especially import-ant to reflect on, now that we're on the verge of transitioning to a new paradigm of automotive transport. The core of this idea is illustrative of what makes human driving so special, and one aspect of what we may stand to lose in an all- (or nearly all-) autonomous era.

Just think about how driving works right now: you get into a vehicle, and using physical motions of your body, you cause it to move, steer, stop, everything. Good drivers know how the car is balanced and gripping and moving on a gut level. They don't assess these things by looking at the instruments and doing bursts of math in their heads, they *feel* it in the same way they feel their body's motion and balance. The same goes for how people who know their cars really well can understand how their cars are performing and operating by feel as well. As you and a car grow used to each other, you begin to learn how it sounds and smells and performs and behaves, and when those behaviors or smells or sounds change, you immediately pick up on that and become aware that something may be amiss.

We can all tell if our car is idling too high, for example, or if there's a change in brake pedal pressure. These are subtle things, but to someone familiar with their own car, they're obvious and can be quite alarming. My very own 1973 Volkswagen Beetle that I drive as of this moment needs a new coil, I think, because I can feel the subtle pulses that mean that at higher speeds/engine revs it's missing in at least one cylinder. I haven't tested all the components yet to confirm this, but I can feel very clearly that something is going on.

Because our physical actions are what control a car and because our bodies directly interpret information about how the car is performing, both on the road and internally, I don't think it's too far a leap to say that automobiles are, fundamentally, prosthetic devices.

Think about how you feel when you get behind the wheel and pedals of a powerful car; you feel powerful yourself, because all those 700 or so insane horsepowers are directly controlled by your very own body; sure, you can get other people and many bags of groceries in there, but the car is really like a mobility suit for you, and the feeling of that can be intoxicating. Riding as a passenger in a powerful car does not give you that same feeling. In fact, many people who love to drive and drive aggressively are the most uncomfortable being *driven* fast and aggressively because they're no longer in control, and it feels wrong, somehow.

Over the decades, humans have adapted remarkably well to

using these motorized prosthetics to move around; we've proven that we're capable of making complex decisions at speeds of well over a mile-a-minute, something that was by no means certain in the earliest days of motoring.

That empowering feeling of driving a car, that satisfaction and excitement, that rush of adrenaline or that unique relaxing feeling from a leisurely drive, these are just not the same when one is a passenger. No one cares about the handling or performance of a city bus beyond the basics of will it get me to work on time and unmangled because as passengers, we're not in a position to enjoy such traits.

Once autonomous cars start to become common, we will be passengers, and our nearly two-century-long experiment in mechanical body enhancement as personal transportation could come to an end. As passengers in autonomous vehicles, we won't have the opportunity any longer to experience the significant joys of being in control of a machine capable of remarkable speed, of handling that satisfies something deep in your gut, or of the ability to traverse terrain far too difficult to walk.

Will future generations look at us and our tedious daily mile-a-minute commutes with disbelief and awe? Driving could become one of those lost arts that nearly everyone once possessed, regarded by generations that grew up with self-driving cars the way we regard people who make their own soap or who know how to butcher a rabbit. There are still people in modern societies who can do those things, but not many, and for those who can, it's a pretty assured way being considered a badass by your friends.

What are we giving up by relinquishing control of our vehicles back to something other than ourselves?

More than we think, and we'll talk about that much more, don't worry.

CHAPTER 2

HOW DID WE GET HERE?

THE GOAL OF A HELL OF A LOT OF HUMAN INVENTIONS IS TO GET out of having to do work. To anyone who has spent any time around humans, this should come as no surprise at all. Work sucks. If I were smart enough to make a machine or program an algorithm to write this book, I would have, no question. The problem is everything I tried just produced long strings of profanity with the occasional phrase "robot cars" stuck in there, and my editors didn't go for it. So here we are.

This mentality is also what led to the development of automobiles. We very quickly realized that walking on our own and hauling our own stuff is awful, just awful, so we forced animals to do it for us, and that worked for centuries, until we just couldn't take staring at a horse or donkey anus from the seat of a wagon anymore, and so the concept of the automobile was born.

I mean, sure, we had to wait until we had a viable power source in the form of, initially, steam power, which was also crucial in finding ways to keep from having to do work. The development of the automobile in general is fascinating, and I originally wanted to write that book but autonomous cars are on everyone's mind right now; I'll get to that history another time.

For the moment, let's just accept that self-propulsion was developed and applied to vehicles, and let's talk about the logical development beyond self-propelled vehicles: self-guided vehicles. The idea of a vehicle that is capable of not just propelling itself, but actually going where you tell it to go, with no further attention or interaction needed, has been around a lot longer than most people would imagine. Some of this may come from a subtle, unspoken desire to replicate animal power; as I said earlier, animal-powered human transportation was at least semiautonomous, and animals on their own are fully autonomous, though you could argue that if you dangle some bacon in front of a dog you have effectively taken control over that autonomy.

There are several levels of autonomy going on here: at the most basic level, any self-propelled vehicle is capable of *some* kind of autonomous travel, even if that just means going forward until it hits something, runs out of fuel, or, for all you flat-earthers reading this, plummets over the side of the Frisbee-shaped planet that you, somehow, believe in.

One step ahead would be to be able to drive according to some set of preprogrammed instructions, sort of like that old '80s toy, the Big Trak. Vehicles of this nature can "play back" a set of instructions: move ahead five units, turn left, pause, that sort of thing. This sort of semiautonomy isn't really autonomy at all, and does not require a machine to have any ability to sense its environment and react accordingly.

Beyond that would be semiautonomous vehicles that rely on sensing very specific external objects, markings, or forces. Think of a car designed to follow a magnetic strip inset into a road, or a robot that uses an optical sensor to follow a line on the ground. These are doing some sensing of their environment, but the environment itself must be tailored to a very specific set of conditions that the machine has been designed to react to.

Then we get to more advanced environmental sensing and reaction; at this level, some degree of computation—first analog, then later digital—is required. Early aircraft autopilot systems used compasses, gyroscopes (a device that uses spinning discs to maintain a reference direction), airspeed indicators, and other equipment to get a sense of where they were going and how fast. While early systems weren't exactly computers, they were sort of like analog almost-computers, reacting to input to determine the environment and act accordingly.

These systems could keep a course, but not actively avoid obstacles; to accomplish this, real machine "vision" had to be developed—and that's a colossal undertaking, the development of which is never-ending—and forms the basis for how modern autonomous vehicle technology works. Modern systems are capable of identifying objects and making best guesses about the nature of the object, and from there extrapolating how it may behave. For instance, something identified as a tree is obviously far more likely to photosynthesize and remain stationary than something that's been determined to be a human on a bicycle, which is far more likely to move, possibly erratically, and then corner you at a party and not shut up about how much better his life is ever since he started riding that damn bike.

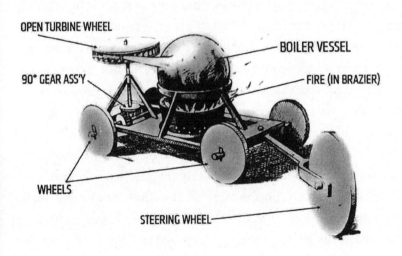

OPEN TURBINE WHEEL

BOILER VESSEL

90° GEAR ASS'Y

FIRE (IN BRAZIER)

WHEELS

STEERING WHEEL

There are, of course, important milestones at all these major stages of development, and unless you can fling this book across the room in time, I'm going to share a lot of these with you right now. In chronological order. So get ready.

The beginning of self-driving vehicle design goes back much further than I think most people would expect; in fact, it goes back further than the automobile itself.

Even if we take 1672 as a starting point for the automobile, when Ferdinand Verbiest made a small (think around two feet long), steam-powered vehicle to amuse the Kangxi Emperor of China, the first self-propelled machine with some crude semblance of autonomy was even earlier. And not just a little earlier; we're talking *two centuries* earlier.

1478: DA VINCI'S CART

You probably won't be surprised to hear that the person responsible for this incredibly early technological wonder is Leonardo da Vinci, a man so far ahead of his time that he routinely had breakfast for dinner. Remember, this is the guy who came up with tanks, helicopters, parachutes, machine guns, scuba gear, and more, all in his spare time when he wasn't painting masterpieces.

It's actually best not to think about his accomplishments too much, because by comparison you'll just feel like a big steaming pile of inadequacy in a funny hat. So let's talk specifically about what da Vinci made. It's usually just called the "self-propelled cart."

The self-propelled cart was arguably the first robot as well, which supports my notion that self-driving vehicles are really just robots we'll be able to ride in. This one though was never intended to carry passengers; it appears to have been intended to be about five and a half feet long by five feet wide by about three feet tall. It was powered by clockwork, which means it was powered by energy stored in springs, and the source of that energy was most likely human muscle. Since it was storing the energy and releasing it on demand, I think that makes it different than a human-powered vehicle like a bicycle; after all, gasoline, if you think about it, is essentially an energy storage system for decomposed dinosaurs and time, but we don't consider cars to be dinosaur-powered.

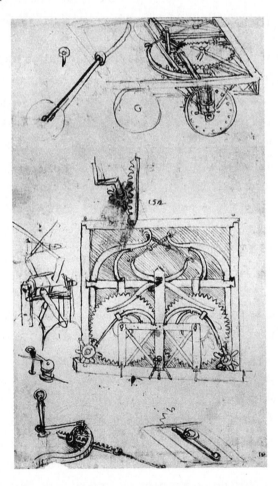

Propulsion was provided by a pair of coil springs housed in drums, and the power from those springs, which diminished as they wound out, was kept steady by the use of a balance wheel, the sort used to keep spring-wound clocks running at a consistent rate. These springs could give the cart an effective range of about 130 feet of travel.

In addition to propelling itself, the cart had a mechanism by which it could be made to turn at preset points on its journey, by placing wooden blocks at locations between gears (some sources describe placing pegs into holes). It seems that only right turns were permitted, but even with that limitation the end result is impressive: this was a programmable machine, capable of executing a stored set of instructions for a very short journey. That, at least in a very simple sense, is an autonomous vehicle.

Da Vinci never actually produced his cart, but a replica based on his original drawings was built in 2004 by Paolo Galluzzi, director of the Institute and Museum of the History of Science in Florence.[2]

1830s–1840s: RAILROADS

There was a pretty significant gap between when da Vinci conceived the cart and when it was produced, which isn't really shocking, since da Vinci's cart was never actually built and, even if it had been, as a fifteenth century Big Trak, it likely wouldn't have had much practical use. This, however, isn't to say that technological development in the automotive arenas wasn't moving ahead; it was, and pretty significantly.

The advent of steam power was, of course, hugely significant in the development of the automobile. Nicolas-Joseph Cugnot had built the first full-size, working automobile in 1769, but since it was incredibly cumbersome and slow, it was doomed to wreck pretty quickly. Developments soon after led Cugnot to create increasingly practical steam-powered automobiles, with purpose-built cars like Trevithick's London Steam Carriage of 1801. He later made others to meet the eventual boom in steam omnibuses in England in the 1830s.

But with the boom came difficulties with powerful horse lobbies that were not willing to lose business to some filthy mechanical upstarts, and this, coupled with generally poor road conditions, forced the early automobile builders to abandon the shoddy network of roads. Instead they laid a network of ideal pathways for their automobiles to traverse. These pathways were extremely low rolling resistance roads, allowing the crude vehicles to carry vast amounts of cargo and passengers; by building a set path for these automobiles, the need to develop steering systems was effectively eliminated, with the pathway handling the steering and navigation through its direction and shape.

We call these pathways "railroads" or "trains."

In many ways, we can think of railroads as one of the earliest semiautonomous vehicle systems, and perhaps to this day still the most common and vast network. A train is an automobile, fundamentally—a self-propelled vehicle, just like the car you drive to the Dairy Queen to do your nightly burnouts in.

There are two major differences between a train and a car: scale and automation. A train is huge compared to a car. Trains, as the earliest automobiles produced in real quantities or really used by the general public, compensated for the crude state of the art by requiring fewer complex mechanisms (the locomotives) and maximizing their use by having them pull long trains of passenger and cargo cars. That's how money was made—locomotives were not going to be the sorts of vehicles sold to every Thomas, Dickomas, or Harryomas in London.

Trains also differ from automobiles, as we understand them today, in that they have one less dimension of operator control than a car. A car's driver can control the speed of the car via the accelerator and brake, and the direction of travel via the steering system. In trains, the operator can also control the speed via the throttle and brake, but directional control is ceded to the machine; in this case, the railroad itself, and its associated switching hardware.

In this sense a train is semiautonomous; an operator is required to control the speed and decide when to stop, but steering is autonomous. This sort of autonomy does not require any processing or understanding of the world on the part of the vehicle itself; the network the vehicle operates within handles that. A railroad is like a vast machine unto itself, with the ability to control multiple vehicles via increasingly complex switching and related hardware. These systems, in some form, were in place even by the mid-1800s.

Railroads were humanity's first successful deployment of a semiautonomous vehicle, and it remains a staggering success.

1866: WHITEHEAD TORPEDO

It's not surprising that war provided the impetus to develop the first semiautonomous vehicle capable of reacting to its environment. I guess it's a little disappointing, but, come on, we know the story; nothing spurs humans on quite as well as figuring out new and exciting ways to blow one another up. I'm not even going to pretend to moralize here, because we all know this is true.

This first vehicle capable of sensing and reacting to its environment wasn't a land vehicle, and it couldn't carry people, just cargo, and that cargo was limited to explosives designed to blow up boats. The vehicle I'm talking about is a torpedo. Back when these were first developed, they were even called "automobile torpedoes."[3] The formal name was the Whitehead Torpedo, a name that sounds like some awful skin-care tool sold in the late 1980s on late-night television. While the basic idea was conceived by others, it was English engineer Robert Whitehead who eventually perfected the design and put it into production. Initially, the torpedo (named for the fish/ray that likes to shock its prey) was just a little unmanned boat that could be launched along the surface of the water to hit an enemy boat, detonate, and—hopefully for the launcher—sink the enemy boat.

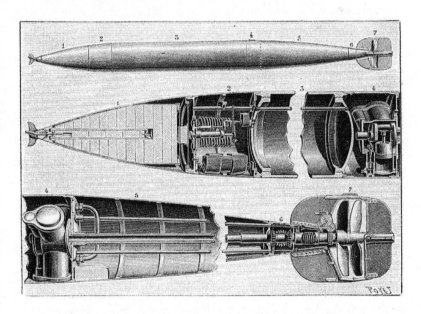

Whitehead added some crucial innovations to the torpedo, and those innovations are what made it the first environment-reactive vehicle: it could keep to a constant, set depth under the surface and it could stay on a fixed course toward its target. Together, these were the makings of the first, crude guidance system, and the first time any inanimate object could really control its direction and compensate to maintain it, even with environmental inputs acting upon it.

To do this, Whitehead installed two pieces of equipment in the torpedo: a horizontal rudder controlled by pendulum balance (to maintain depth) and a hydrostatic valve (a one-way, pressure-relieving valve), and a gyroscope system driving a vertical rudder to keep it on course. These systems allowed the torpedo to control its path on two dimensions, with the third (forward travel) dimension provided by a three-cylinder radial-compressed air engine.

The pendulum-and-hydrostat control of depth is ingenious. A hydrostat senses the depth, but does not control the horizontal rudder directly; if it did the torpedo would oscillate around the desired depth without ever really settling. The pendulum swings based on the pitch of the torpedo, and is connected to the rudder control in such a way that it can dampen the oscillations, providing much steadier control over the depth of the torpedo. The pendulum-and-hydrostat device was such a big deal at the time that it was called the "Whitehead Secret,"[4] and the same fundamental design was used all the way up to World War II.

The gyroscopic control for azimuth/yaw control came in 1895—prior to that, the azimuth (you know, direction, basically) was set with vanes by hand. The gyroscope in the Whitehead torpedo was spun up via a spring, and acted on the vertical rudder via gimbals. This kept the torpedo on a straight, direct path regardless of whatever forces (such as ocean currents) were acting upon it.

1925: HOUDINA AMERICAN WONDER

The first conventional automobile to be driven without a person at the wheel was developed in 1925, but it's really sort of a cheat. It wasn't an autonomous car, but rather a remotely controlled car, so it was still driven by a human even though the human wasn't inside the car.

Pictures of the car show it labeled as a "1926 Chandler," which is sort of confusing, since it appears to have been demonstrated and in operation since 1925. The car, nicknamed the "American Wonder," was built by an electrical engineer named Francis P. Houdina.

The way it worked was pretty straightforward: the car had a kite-shaped receiving antenna mounted on the tonneau, and electric motors under radio control to actuate the controls. It's not entirely clear how many of the car's controls were controlled by the motors or how the mechanisms worked. We do know the steering seems to have been accomplished with a belt or similar device around the steering shaft itself, because a poor grip on the steering column caused some excitement during a demonstration drive in New York in the 1920s.[5]

Here's how the *New York Times* described it:

A loose housing around the shaft to the steering wheel in the radio car caused the uncertain course as the procession got underway. As John Alexander of the Houdina Company, riding in the second car, applied the radio waves, the directing apparatus attached to the shaft in the other automobile failed to grasp it properly.

As a result the radio car careened from left to right, down
Broadway, around Columbus Circle, and south on Fifth
Avenue, almost running down two trucks and a milk wagon,
which took to the curbs for safety. At Forty-seventh Street
Houdina lunged for the steering wheel but could not prevent
the car from crashing into the fender of an automobile filled
with camera men. It was at Forty-third Street that a crash
into a fire engine was barely averted. The police advised
Houdina to postpone his experiments, but after the car had
been driven up Broadway, it was once more operated by
radio along Central Park drives.[6]

It seems that, at a minimum, there were mechanisms for
steering, starting the car, actuating the throttle pedal and brake
pedal, and perhaps clutch and shifting. It's possible they just left
it in first or maybe second gear, though I think they'd need some
degree of clutch actuation.

The thing seemed to work generally well enough for a proof
of concept, and in the overall scope of autonomous vehicles the
American Wonder proved that motors, servos (automatic devices

with some form of error-sensing and correction), and similar mechanisms could be used to actuate conventional car controls in place of actual human limbs and hands. If we replace those radio signals from a human with signals from onboard cameras, sensors, and computers, you've effectively got the basics of how modern autonomous vehicles are built.

One fascinating footnote to this has to do with the inventor's name: Houdina. As you probably already noticed, that name is an awful lot like Houdini, as in Harry Houdini, the famous illusionist and escape artist. Houdini was not the sort of person to take guff of any kind, ever, and he felt that Houdina was deliberately using a name that sounded like Houdini for the name of his company, Houdina Radio Control Co. Houdini didn't seem to care that the man's name was, in fact, Houdina, and was convinced it was all just some dirty ploy to capitalize on Houdini's success and name recognition. Guys who escape from chains underwater don't usually write tersely worded letters, and Houdini instead opted for the more direct method of going to Houdina's office and trashing the place.

Houdini wrecked some furniture and an electric chandelier, and pitched what must have been a very exciting fit. Houdini was summoned to court regarding the incident, but no one from Houdina showed up, so Houdini got away scot-free with the perfect crime of chandelier damage.

1933: MECHANICAL MIKE AUTOPILOT

Even though at the moment autonomous control for cars is the hot topic of (admittedly geeky) conversation, it's worth remembering that airplanes have been flying themselves, more or less, for *decades*. At first glance this may seem counterintuitive—aren't aircrafts dramatically more complex than cars? How do they routinely employ self-piloting systems that automobile makers are still struggling with?

The answer is pretty evident when you think about it for even a slight moment, and chances are most of you already realized it while reading that last sentence. It mostly has to do with this one indisputable fact: the sky is really big and really empty. Obstacle avoidance isn't really that pressing a concern in the air. The chances of a cyclist pulling out unexpectedly in front of you in the sky are exceedingly remote; even if one did, they'd have much bigger things to worry about than getting hit by a passing airplane.

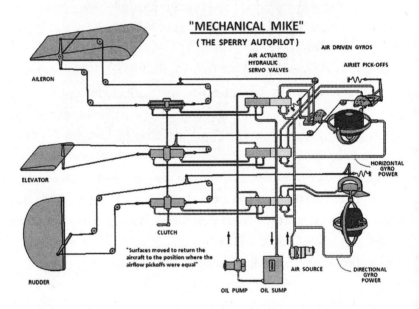

"MECHANICAL MIKE"
(THE SPERRY AUTOPILOT)

It's sort of counterintuitive, but the air is a pretty forgiving place in which to develop autonomous piloting systems, even when accounting for the fact that if anything goes wrong the equivalent process of pulling over to the side of the road ends in a fireball on the ground. The sky is vast and empty, and that's why fairly crude devices like the Mechanical Mike Autopilot, the first really practical aircraft autopilot, were so successful.

Autopilots like Mechanical Mike and most of the ones that followed are self-piloting in that they can maintain a set course and heading (that is, the compass direction an aircraft is pointed) and altitude, but they're not concerned with any real obstacle avoidance, unless you count the ground, which is, admittedly, a pretty significant obstacle.

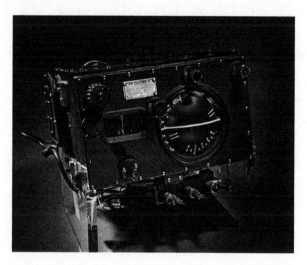

The first aircraft autopilot system was developed by Sperry in 1912, and in many ways wasn't that different from the secret of Whitehead's torpedo. The original autopilot, known as a "gyroscopic stabilizer apparatus,"[7] was composed of a pair of gyroscopes, one connected to the heading indicator and one to the attitude indicator and controlling, via hydraulics, the airplane's elevators and rudder. This early autopilot allowed a plane to fly level on a particular compass heading, freeing the pilot from many tedious hours of constant attention.

The first really significant use of an autopilot system came in 1931 when Wiley Post set a record for flying his Lockheed Vega airplane around the world in less than eight days. Post's Vega had a more developed (than that original 1931 system, at least) Sperry Autopilot installed, which Post nicknamed "Mechanical Mike."

For what it did, Mechanical Mike was surprisingly small, being only about 9 inches by 10 inches by 15 inches.[8] The box housed two air-driven, 15,000 rpm (revolutions per minute) gyroscopes, one for azimuth/direction and one for lateral control of the airplane. Air-actuated servo valves connected to the gyroscopes hydraulically controlled the aileron, elevator, and rudder of the plane, giving full three-axis control of the aircraft.

Mechanical Mike required no electrical power, being entirely pneumatic, and only weighed seventy pounds. These were significant advantages over other autopilot systems, and Mike proved to be remarkably reliable. Mechanical Mike and other early gyro-based autopilot systems are significant in the development of autonomous vehicles because they represent the very first time a fully mechanical vehicular control system was trusted enough to transport passengers. The wide use of the Mechanical Mike following Post's trip marked the first mass deployment of a situationally/environmentally reactive autonomous vehicle system, and, with nearly every major aircraft today employing a much more advanced version, represents the most common autonomous vehicle fleet currently in use on Earth.

WORLD WAR II: PROJECT PIGEON

The desire to have an autonomous system that could not just react to basic input about the environment—like speed and heading— but could actually sense, track, follow, or aim at a particular target, has been around since long before we had the means to produce machines capable of such tasks. That's why things like

Project Pigeon came into existence; sometimes, we didn't want to wait around until we developed such machines, and instead borrowed the equipment we needed from nature. Even if that equipment was pigeons.

An autonomous vehicle that can track an object and make turns or otherwise modify its course to track an object is, naturally, highly useful in warfare, where so much is concerned with sending moving things along very fast to chase down and slam into other moving things. To pull this off, you need some sort of visual tracking and processing system, which pigeons already have.

The famous behaviorist B. F. Skinner realized this, and during World War II employed specially trained pigeons to steer a missile at his intended target. Skinner developed a nose cone for a missile that contained three windows, each one with a pigeon looking out of it. These pigeons had been trained via operant conditioning[9] (basically, rewarded when a desired behavior was performed and punished when not) to peck at a target seen through the window. By pecking on the window that contained the target, the pigeon actuated controls that allowed the missile to adjust its course and keep the target centered.

The system was said to have performed well in simulations but Skinner complained that no one would take him seriously.[10] There's a lot of reluctance for people to entrust weapons of war to pigeons, a prejudice that still holds even to this day.

1940: AUTOMATIC TRANSMISSION

While not strictly a self-driving innovation, the development of the automatic transmission should be considered a step on the way to autonomy for automobiles. It was even referred to as "self-drive" in many early contemporary descriptions, because to people used to the near-constant shifting of gears, it felt like the car itself was doing a lot more work.

And the truth is it was. Determining the proper gear ratio for the needs of the car wasn't trivial, and the automatic transmission freed the driver to focus on the more fundamental tasks of driving, speed modulation, and directional control.

First developed by General Motors (GM) (specifically its Cadillac and Oldsmobile divisions), the first commercially available automatic transmission was GM's Hydra-Matic. Developed in 1939, the Hydra-Matic was the result of a number of earlier research efforts to make a self-shifting transmission, and used a combination of a planetary gearbox—a transmission type that uses a central sun gear and orbiting planet gears (you see why it's called that, right?)—and a novel fluid coupling to achieve its goal. While heavy and complex, it was a success, and, for the American market in particular, regarding automatic transmissions, the industry never looked back.

1945: CRUISE CONTROL

Incredibly, automotive cruise control was invented by someone who didn't even drive. In fact, it was invented by a blind man, Ralph Teetor, who came up with the idea after being annoyed at how bad his lawyer drove.[11] Teetor was riding with his lawyer and having a conversation with him, and felt the car slowing whenever the lawyer was speaking, and accelerating when he was listening.

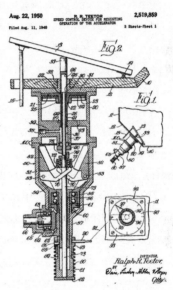

Nauseous but determined, Teetor wanted to find a way to keep a car's throttle constant and free from the human driver's fluctuation, and so developed cruise control, the first real semi-autonomous assist device for cars.

Engine speed governors designed to keep an engine running at a constant speed have existed for over a century—you know those two spinning balls you sometimes see on old steam engines? That was a centrifugal governor, and also the origin of the phrase "balls to the wall," since at maximum speed, those balls would be flung out nearly horizontally, or, you know, to the wall.

Balls or not, just keeping the engine at a constant speed really isn't enough for a car cruise control system. Teetor computed the car's actual speed based on how fast the driveshaft was rotating, and then used a bidirectional electric motor connected to the carburetor's throttle to adjust the position of the throttle to keep the desired speed constant.[12]

Chrysler was the first to market the system, which was variously known as Speedostat or Auto-Pilot (foreshadowing Tesla's name for their semiautonomous driving system), and later Chrysler came up with the name "Cruise Control," which eventually stuck.

The cruise control innovation is important because this was the first taste most drivers had of any sort of actual driving automation. Sure, automatic transmissions—and before that, automatic spark advance and oiling and so on—took over many of the functions of the operation of a car that used to be manual, but those, including gear shifting, were less about the actual piloting of the vehicle itself and more about the technical requirements needed to get the car to drive at all.

Cruise control, though, was clearly different, in that it took one of the primary tasks of driving—speed control—away from the human driver and placed it under control of the machine. Sure, this was the simplest form of control, with no ability to independently sense its environment, but it was a start. Modern dynamic-cruise systems use radar to keep a set distance away from the car in front, and can brake automatically if the system determines it's approaching an object in front of it too rapidly.

All of this is thanks to one blind man and his lawyer—who couldn't drive and talk at the same time.

1956: GM FIREBIRD II CONCEPT CAR

The first time a major carmaker presented the idea of real autonomy to the public was in 1956, when GM showed their concept car, the Firebird II. As the name implies, this wasn't the first Firebird concept car, but it was the first of the Firebird concept cars to suggest a then still-fictitious world of automated driving.

The Firebird concept cars were designed by Harley Earl and were heavily influenced by jet fighter designs. The Firebirds were showcases for cutting-edge technology, and as such employed exotic gas-turbine drivetrains, and cathode-ray tubes (CRTs—you know, the old big tube kind) in its television- and camera-based rearview systems.

More important for our purposes, though, is the automatic driving system that GM imagined for the car. This wasn't just a concept for the car itself, but rather an entire network for automated vehicles, including specially prepared roadways with integrated "conductor strips" and control towers in the "Autoway Safety Zone," the name given to the automated highway system in the film.

GM describes the system in a brochure published for the Firebird II's introduction at GM's Motorama show at the 1956 World's Fair:

> This amazing concept places control of the motorcar in
> the hands of an "electronic brain"—actually releasing the
> driver from the wheel. . . . These include a Dashboard View
> screen which has two panels. The left panel is for "internal
> communication" between car and driver (information he
> would normally receive from visible instruments as to fuel
> supply, engine operation and temperature). It also reveals

a radar pattern when he guides the car onto the electronic control-strip for automatic steering. . . .

Extending from the two engine air scoops on each side of the nose of the car are probes or antennas which pick up wave impulses from the conductor strip in the center of the control lane.[13]

Of course, none of these things actually worked, but it is interesting to see how the very sticky problems of computer vision could be avoided if fully autonomous operation is limited to areas where an infrastructure has been built to guide the cars.

No mention is made of obstacle avoidance or anything like that; presumably, it is the job of the singing gentlemen in the control towers to make sure everything is running smoothly and to warn drivers to stop if they're approaching a broken-down vehicle or a coyote on the road or something else they don't want to barrel through. I'm not confident that would have worked out so hot.

Really, this sort of system is more like a hybrid of a tram or trolley car type of vehicle combined with a conventional car. On roads with the proper control towers and guidance strips, the car cedes control to the roadway network, much like a tram or train. On roads without the necessary hardware, you're just driving a normal car, even if it is powered by a kerosene-fueled turbine and has a clear bubble dome.

Because of the very significant infrastructure investments required in such a system, nearly all modern self-driving technology and research is designed to work without outside, physical infrastructure help, which is, of course, a much more difficult task.

One more thing about the Firebird concept, and, specifically, the film GM made to promote it. The film shows a family on a

road trip in their Firebird, enjoying all of the considerable com-
forts of 1950s-envisioned 1976 future life. At one point, based
on the recommendation from the guy in the control tower who
assures them that the "hostess is a dream," they contact a nearby
hotel on their dashboard video screen, where they see a lovely
woman who sings the praises of the hotel.

At one point in the song she tells them about the restaurant
at the hotel, and sings this line: *"Our pre-digested food is
cooked by infra-red!"* I think it's safe to say that what people
are likely to find appealing has changed *dramatically* from what
it was in the 1950s.

1957: RCA LABS AND THE STATE OF NEBRASKA'S EXPERIMENTAL HIGHWAY

Well, "highway" is a bit generous, since this was just a four
hundred-foot stretch of road, but you get the idea. A Nebraska
state traffic engineer named Leland Hancock was very taken by
the idea of automatic control of cars on the highway to help

combat driver error and fatigue, and to prevent accidents, and was determined to get others interested in the idea. Thanks to a lot of determined letter writing, he was able to get researchers at the RCA Corporation to work with him. Together they arranged to lay coils of wire at the intersection of US Route 77 and Nebraska Highway 2 as the roads were being built.[14]

On October 10, 1957, the development team carried out a test, witnessed by eighty-three people. Using a 1957 Chevrolet Bel Air with antenna coils mounted on the bumper, a special meter in the car, and a partially obstructed windshield, a driver was able to drive on the road and follow its course by watching the deviation of the meter's needle. If the car got too close to a car in front of it, an alarm sounded, ringing a bell and flashing a light until the car slowed down enough to open the distance to an acceptable level.

While the car wasn't doing the driving, relying on a person to actuate the controls, it did replace the human driver's need for vision; if researchers had chosen to, the system could have been rigged to actuate the car's controls directly.

1960: UK TRANSPORT AND ROAD RESEARCH LABORATORY'S EXPERIMENTAL FOUR MILES OF M4

In an experiment quite similar to the RCA/State of Nebraska one, the Transport and Road Research Laboratory (TRRL) of Crowthorne, Berkshire, in the United Kingdom buried a four-mile-long length of cable beneath the stretch of the M4 motorway between Slough and Reading.[15] The cable was laid, as in Nebraska, while the road was being built, and experiments were performed on the stretch of road before it was opened to the public.

Stills from Key to the Future, *GM's film about the Firebird II, showing its automatic driving features and the control towers of the Autoway Safety Authority. This is worth watching because there's lots of singing involved.*[16]

An early 1960s experiment using a Standard Vanguard was quite similar to the Nebraska/RCA one: the car had its windscreen obscured with cardboard, and information about steering was conveyed via an indicator mounted to the dashboard.

A more sophisticated experiment followed, this time using a Citroën ID19. The Citroën had an advanced hydropneumatic suspension system, and this high-pressure hydraulic system was used to drive actuators for the steering, brakes, and throttle. Using the system and receiving guidance information from the embedded cable, the Citroën was able to drive completely independently, and was tested at speeds up to 80 mph, as well as in ice and snow. It performed remarkably well in all the tests. But while results were promising, it was determined that implementation would be too costly based on (in hindsight, woefully conservative) estimates of future traffic growth; the TRRL was ordered to stop development and research on the project.

1961 TO 1979: THE STANFORD CART

While not designed to actually carry passengers and not exactly a car, as such, the humble Stanford Cart could be considered the real start of modern autonomous car technology. The Stanford Cart was just what it sounds like: a little, wheeled cart, sort of a small table with wheels, developed at Stanford University.

The cart was originally a test platform to study how to possibly remotely control a lunar rover from Earth. The Jet Propulsion Laboratory was developing a lunar rover that would be controlled by radio signals from Earth, and information sent back from a television camera. Grad student James Adams built the cart in 1961 to simulate the rover, taking into account the 2.5-second delay in the signals going to and coming back from the moon. He was able to prove that with the delay, the rover would not be reliably controllable at speeds over 0.2 mph, which is, as you can guess, really, really slow.[17]

To overcome this, experiments began to attempt to give the cart its own ability to "see" its environment, detect obstacles, and take steps to avoid them. This was the birth of nearly all computer vision systems employed by autonomous vehicles (and, really, any robot that uses some manner of camera-based synthetic vision) today.

By 1964 the cart had been re-outfitted with a low-power television transmitter that broadcast TV signals to a PDP-6[18] computer to process the images. With this setup, which I'm dramatically simplifying here, the cart was able to visually follow a high-contrast white line on the road at about 0.8 mph. This was a big deal, as it represented real computer vision controlling a moving machine, even if it was quite crude.

Development continued on the cart with new researchers and students re-outfitting the cart as new ideas and technologies became available. In 1977, the cart was upgraded with faster processors, an independently mobile camera, and four-wheel steering; in this configuration it was able to drive around obstacles in a controlled environment.

Sure, it only moved three feet at a time and then had to pause to figure out what it was looking at, but this was a gigantic leap in robotics. The cart was eventually able to navigate a chair-filled room in about five hours, and while it may be tempting to laugh at that idea now, I can think of plenty of times I've not been able to navigate a chair-filled room without running into half the chairs and looking like an idiot.

1977: TSUKUBA MECHANICAL ENGINEERING LAB, JAPAN

Arguably the first fully autonomous, computer-vision-controlled car was shown in 1977 by the Tsukuba Mechanical Engineering Lab, in Japan. The project, headed by Sadayuki Tsugawa,[19] modified a full-size car to follow special white road markings and was able to drive at speeds of nearly 20 mph. While still essentially a follower of specially contrived external visual guides, the fact that this technology was implemented in a full-size car driving at a reasonable speed (compared to, say, the Stanford Cart) and using computer-interpreted visual information made this a significant milestone.

1980S: ERNST DICKMANNS: THE MAN WHO MADE CARS SEE

If the overall concept of vehicles driven via true computer "vision" can be said to have a father, that father would have a German accent and a hilarious last name: Dickmanns. Ernst Dickmanns was the father of a particular set of Mercedes-Benz cars and vans that drove via information captured from cameras and interpreted by some very hardworking computers.

Dickmanns started out working in aerospace, including a stint at NASA, where he researched orbiting spacecraft reentry. By the early 1980s he'd migrated to focusing on the development of machine vision to allow autonomous driving.

Dickmanns's first real application of his research was the result of a partnership with Mercedes-Benz, which was hoping to have something really exciting to unveil for their centenary in 1986: a self-driving car. To achieve this goal, Dickmanns outfitted a Mercedes-Benz L508D T2 van with a lot of computing hardware, cameras, servos, and actuators to operate the van's driving controls.

Since it was 1986, the computing hardware, while state-of-the-art, wasn't really fast enough to process a full visual field captured by the camera in real time—a full one hundred seconds was required to process a full-frame image from the camera. To get around this, there was a master computer, but actual processing of the images from the camera was passed to a parallel-processing system consisting of ten Intel 8086 CPUs—the same ones that powered the IBM PC AT.

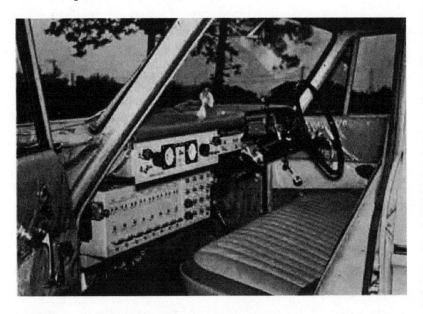

These ten CPUs would only process certain tiny 32 by 32 pixel areas determined to be interesting in some way. The result was that the important parts of the camera's feed (high contrast areas, areas in motion) could be processed much quicker, with only twenty microseconds needed to extract the important and required information from the visual feed.[20]

Dickmanns called his machine vision solution "Dynamic Vision," and considered it a "4-D" approach to machine vision, incorporating both strict processing of the video feed and time delays, as well as predictive behavior information for certain classes of identified objects, things like "spatiotemporal models for motion processes of objects," and more.[21]

Dickmanns's first autonomous van, called the VaMoRs, first drove autonomously on non-public sections of Germany's Autobahn in 1986. By 1987 the vans were on public roads driving at nearly 60 mph; collision avoidance and obstacle detection were implemented, allowing the van to follow a car in a convoy if desired.

Next came the PROMETHEUS Project, a tortured acronym for PROgraMme for a European Traffic of Highest Efficiency and Unprecedented Safety. These were also Mercedes-Benz programs, but now much classier, since the components had been miniaturized enough to move them from a van to an S-Class Mercedes.

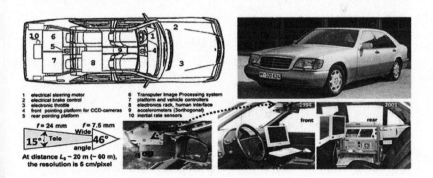

These cars (known as VaMPs) managed speeds of up to 115 mph on the Autobahn, and were even able to pass slower cars. Eventually they were able to drive over 1,000 miles with minimal human intervention. They benefitted from much upgraded hardware from the van era, at first using transputers (a parallel processing architecture) and then later PowerPC 601 chips, which you geeks may remember as the architecture that Apple Macintoshes used after the chips from the Motorola 68000 family and before Intel chips.

Dickmanns's work was hugely influential, and laid the template for AVs to follow.

2004: THE DARPA GRAND CHALLENGE

If there was one final crucible that truly made autonomous vehicles a viable, achievable possibility, it had to be the DARPA Grand Challenge.

The Grand Challenge was a project run by the US Department of Defense's Defense Advanced Research Projects Agency, which was authorized by Congress to offer one million dollars of prize money to the first team that could build an autonomous vehicle capable of traversing a 150-mile-long route in the Mojave Desert that follows the path of Interstate 15.

The course has many turns, obstacles, switchbacks, roads near precipitous drop-offs, and probably roadrunners being pursued by coyotes on Acme brand rocket skates. It's not an easy course, a fact that was made very clear in 2004, when the first Grand Challenge was run. There was no winner that year; the best showing was by Carnegie Mellon's Sandstorm, a converted Humvee that made it all of 7.32 miles before getting stuck on a rock.

Since there was no winner in 2004, the Grand Challenge was held again in 2005. This time five vehicles successfully completed the course, with the winner being Stanford University's Stanley, a converted Volkswagen Touareg. Stanley navigated the route in six hours and fifty-four minutes.

In 2007, the Grand Challenge was back, this time with an "urban challenge" designed to replicate the challenging environment of city driving. This challenge required that the cars meet all traffic laws and interact with one another legally,[22] following regulations and conventions such as moving in the proper order when four cars meet at a four-way stop sign. That year Carnegie Mellon's team won with a very modified Chevy Tahoe named "Boss."[23]

The Grand Challenge victory in 2004 can be seen as the moment that autonomous vehicles graduated from the laboratory and went into the real world. The Grand Challenge victory by Stanford was one of the first times the mainstream public became aware of all the research into autonomous vehicles, and the first time many people and businesses saw that autonomous cars weren't just some weird, bubble-top painting from a *Popular Mechanics* they sort of remembered seeing on their uncle's desk before he went to jail for that thing, but were actually becoming not just technologically possible, but something they might even own in the future.

Since we're talking about timelines, it's worth mentioning that, to be honest, I think this book is somewhat premature. I don't think the coming age of autonomous cars is quite as close around the corner as you may be led to believe when some Tesla owner corners you in the gym locker room and won't shut up, even though the dude is dripping wet and very naked.

AVs are robots, and they're incredibly complex, and the real-world testing needed to actually figure out how they will work and interact in the chaotic, messy reality we live in has just barely started. They're coming, yes, but likely not as quickly—at least in significant numbers—as many people seem to think.

That's fine, though, since we have an awful lot to figure out before they get here. Let's start by looking at how these things will work.

CHAPTER 3

HOW DO THEY WORK, ANYWAY?

IF WE'RE GOING TO TALK AND THINK ABOUT AUTONOMOUS CARS, self-driving cars, robo-cars, drive-o-droids or whatever the hell we want to call these things, we should get a sense of exactly what they do and how they do it. Those of us who drive may want to think back to a time before driving was all ingrained muscle memory, something we now do almost automatically, except when we're trying, like idiots, to drive while texting someone something super important about this hilarious drunken fight we think we saw behind a supermarket.

Let's take a moment to break down what the act of driving is: controlling a moving object significantly larger than yourself through a complex environment full of other moving objects; controlling the rate of its motion, including entirely stopping the motion; controlling the direction of that motion; and adjusting speed and direction in response to environmental conditions, like road surface friction, weather, and visibility.

The more you think about it, the more you realize it's pretty incredible that we can drive at all. For one thing, we're asking our bodies to act and react at far quicker response times than we were designed to do. Sure, many of the traits that make it possible for humans to drive at all are the result of millions of years of evolution as predators and hunters, which gave us such handy-for-driving traits as forward-facing binocular vision, the ability to track moving objects while we're in motion (handy for chucking spears at mammoths and bison), and the ability to anticipate the motion of other moving things.

Being bipedal is nice, too, since it has conditioned us to be able to use our hands for complex tasks even while in motion. Of course, the sort of motion we were "designed" for was more on the range of a good jog, maybe around 5 mph or so. We routinely drive at speeds *ten times* that speed, and seem to manage it fairly well. Still, what we're doing by driving is way out of warranty, though, to be fair, everything we've been doing since we figured out how to build fires is warranty-voiding behavior, so that barely matters.

When we drive, we rely on our vision, primarily to get a sense of the world around us, but there's a lot more going on. Years and years of experience being a human living in the world have prepared us to understand the things we see on the road, and to anticipate predictive behaviors for the things we see. When we see a cyclist riding alongside the road, we have a good general sense of what sort of speed to expect compared to something similar, like a motorcycle.

We can look at a pedestrian on a street corner and tell, thanks to an innate understanding of human behavior, if that person is paying attention to their surroundings or not, and we can adjust our focus on the person accordingly. We can often tell if the person sees us, in the car, and can use that information to decide how best to approach, or even to alert the person to our presence by blowing the horn, which may or may not play a ridiculous version of "Dixie."

We, as drivers, are able to compensate for gaps in what we see or expect, thanks to our experiences. If we're on a road that goes from having clearly marked lane divider lines to a section where none of the lines are visible, we can deal with that because we know roughly where the lanes should be and can continue based on what the last visible section of lane markings told us.

We rely on hearing as well; think about how many times you've prepared yourself to be passed by a loud motorcycle or muscle car. We know they're approaching long before we see them, and we can tell what side they're coming from. The same goes for approaching sirens of emergency vehicles. Think also about how you react to hearing a horn honk somewhere around you, and the heightened state of awareness that puts you in.

There's a lot of driving that we do by feel as well. By the way the car's weight shifts, a driver determine if a curve can be taken safely and sustainably. We can feel how the tires grip or don't grip a given road surface, and adjust our driving accordingly, or, at least, we should. We can even feel how the car brakes, accelerates, and idles, to get a sense of how the car is performing, technically, allowing a good driver to maintain a general awareness of the condition of the car, and knowing what can, can't, or shouldn't be done with it.

Hell, we even use our sense of smell to determine the state of our vehicle, as anyone who has ever burned out a clutch can tell you. Our sense of balance also comes into play, helping us react if we almost tip over a top-heavy SUV.

What I'm getting at here is that driving is very much a full-sensory experience. The act of driving is not a purely visual one as many assume: it involves your entire body, which is why building a machine that drives itself is such a complicated, involved ordeal.

Machines don't have many of the innate abilities that humans have and use while driving, so to make a car able to drive itself, one pretty much has to start from scratch. Machines, though, have some pretty significant advantages to help them out, like the inability to feel fatigue, remarkable speed, and the ability to be interfaced directly with the machinery of the car and receive data from other machines, as well as other non-human qualities.

Most modern autonomous vehicles rely on the same basic set of technologies and general procedures to work, so let's walk through the basic tool kit that allows an inert machine to drive itself around just like your deadbeat cousin does every day.

The equipment on an autonomous vehicle that differs from that of a human-piloted vehicle can be classified into two categories: sensory equipment and mechanical actuators. The actuators convey the driving system's decisions to the physical part of the car, and can be very tightly integrated with the car itself. Motors that are used to assist with steering for human drivers can also be used to turn the steering rack. Since most modern cars are drive-by-wire—that is, not requiring a physical cable from the gas pedal to a butterfly valve in the fuel system, but instead sending electronic signals from the pedal to increase or decrease throttle—having something other than a human controlling the functions of a car is pretty straightforward. This part of the equation is by far the easiest part.

The hard part is getting the car to perceive and understand the ever-changing, ever-moving world around it. That's where the sensory equipment comes in. Here's what that equipment usually consists of:

ULTRASONIC SENSORS

You know those little round button-like things you see on some cars' bumpers? Those are ultrasonic sensors, and they're most often used as parking-assist sensors, since they're good at telling what's close to you, at low speeds. They bounce ultrasonic sound waves off objects to determine how close you are to them.

These don't have much use in fully autonomous vehicles, but they do help a car understand its environment, so they're worth a mention. Automatic parallel parking systems use them, so there are *some* autonomous-driving/parking contexts where they're used.

We can't hear the pulses they make, since those pulses, while loud, tend to be between 40 kHz and 48 kHz (or higher, in newer sensors).[24] Humans generally don't hear above 20 kHz. Dogs, cats, and bats, though, can hear these pulses, which must be pretty annoying.

CAMERAS

Vision, is, of course, the most important sense we use when driving, so most self-driving machines need a way to replicate it. Modern technology is capable of making some very small and high-resolution camera systems, and modern cars already are getting pretty laden with cameras, even if they don't have any pretentious interest in driving themselves.

The rearview camera is probably the one most people are familiar with, although its output is designed for human drivers to see—it's a dumb camera. Other cameras, usually mounted just above the inside rearview mirror in the top-center of the windshield, are used for lane-departure systems, in which computers run software that analyzes each frame of video to identify the lines painted on a highway and make sure the car stays inside them. These cameras may also be used with emergency braking systems and traffic sign identification. All of these examples require camera systems with some degree of artificial intelligence, since they're attempting to make some sense out of the images they capture.

"Sense" is a bit of an anthropomorphizing term, of course: they're really just analyzing frames of video for a very specific set of criteria, and acting on that criteria in very defined ways.

Most autonomous vehicle camera systems use two cameras to provide binocular vision for real depth perception. While these cameras are good, they're not usually as good as the one in, say, your phone. Most tend to be between 1 and 2 megapixels,[25] which means they're imaging the world at a resolution of about 1600 by 1200 pixels. Not bad, but much less than human vision. Still, this seems to be precise enough to accurately perceive the environment, and to process images quickly enough for driving.

Really, it's not about image quality or color saturation or any of the sorts of criteria we normally use when we evaluate cameras for our own use. For driving a car, you want fast image acquisition—the more frames per second you can capture and evaluate, the quicker the car's reaction time will be.

When processing images from the camera, the car's artificial vision system has to look out for and identify a number of things:

- Road markings

- Road boundaries

- Other vehicles

- Cyclists, pedestrians, pets, discarded mattresses, and anything else in the road that is not a vehicle

- Street signs, traffic signs, traffic signals

- Other cars' signal lamps[26]

To identify these objects and people, the camera systems must figure out which pixels in the image represent background and which represent things that need to be paid attention to. Humans can do this instinctively, but a machine doesn't inherently understand that a 1600 by 1200 matrix of colored pixels that we see as a Porsche 356 parked in front of the burned remains of a Carl's Jr. is actually a vehicle parked in front of a subpar fast-food restaurant that fell victim to a grease fire.

To get a computer to understand what it's seeing through its cameras, a number of different methods have to be employed. Objects are identified as separate from their surroundings via algorithms and processes like edge detection, which is a complex and math-intensive way for a computer to look at a given image and find where there are boundaries between areas, usually based on differences in image brightness between regions of pixels.

As you can imagine, this process is nontrivial,[27] since any given scene viewed through a camera is full of such things as gradients of color and shade, shadows, bright spots, and confusing boundaries. But complicated math that looks like this

$$f(x) = \frac{I_r - I_l}{2} \left(\operatorname{erf}\left(\frac{x}{\sqrt{2}\sigma} \right) + 1 \right) + I_l.$$

is precisely the sort of thing computers are good at, so, generally, this process works quite well.

Once individual objects are separated from their backgrounds, they need to be identified. Size and proportion are big factors in this, since most cars are—very roughly—similarly sized and proportioned, as are most people or cyclists and so on. Things that are large 12 by 5 by 6 foot rectangles are likely cars, narrower things that are positioned like a book with its spine out and its body behind it are probably bicycles or motorcycles, and tall oblongs that move around are probably people or magic walking cacti.

While most autonomous systems are pretty good at identifying cars, people, and bikes, they're still pretty stupid compared to humans. For example, while we would never mistake this[28] for a real car:

it's close enough for an autonomous car to misread it. Good object identification is accomplished via lots of training and machine learning with thousands and thousands of pre-categorized example images, and while it's extremely impressive and works, it can be fooled in troubling ways. For example, it's hard to tell the difference between a picture of a bicycle and a real bicycle, especially if the bike is an image on the back of a moving car, allowing the bike image to move as would be expected by the computer.

This image of a van with a printed image of people on bicycles is from an article in MIT Technology Review,[29] and it highlights the biggest issue with cameras and image identification systems: they're easy to fool, end, even if we're not talking about deliberate trickery, they can get confused. The solution to this is to not just to rely on cameras, but to use cameras as part of a larger suite of sensors.

There are lots of good reasons to have many different world-sensing options as well; for example, being able to "see" what's going on even in conditions where visibility is limited. Darkness is a factor, of course, but so is bad weather. We've all been driving and gotten caught in torrential rains that render the view out our windshield into something that looks like what you see if you attempt to view the world through a nice, cold gin and tonic. You can't really see, and a computer wouldn't be able to either. But, other systems, like radar or lidar, both of which we'll get to soon here, may not be as affected.

As for the number of cameras to be used, at minimum an autonomous vehicle would need a pair of stereo cameras facing forward, though the addition of rear and side cameras to help get as close to a 360-degree view as possible would be ideal.

RADAR

Cameras give a good overall view of the environment around the car, but being able to use that view to navigate real three-dimensional space requires some complicated math and processing. Radar systems are used to help the car understand how far it is from other cars and objects around it.

Radar systems are already fairly common in cars today, as they form the basis of dynamic cruise control systems. Adaptive cruise is a form of semiautonomy, in which the car drives at a set speed, like normal cruise control, but uses the radar emitter to determine its distance from the car in front of it and adjust its speed to maintain the desired distance.

You can usually tell if a car has a radar emitter/receiver by looking at its front. If you see a strange, shiny flat panel masquerading as a piece of the grille, or if the front badge appears to be "printed" on a solid, shiny black panel, then you can safely assume a radar transceiver is mounted there.

Radar data doesn't attempt to give the full view of a camera-based vision system, but a car equipped with it can more reliably process distance information and is more tolerant of darkness and inclement weather conditions that could confuse or impair a camera-based system than a car without it is.

LIDAR

In some ways, lidar is one of the most controversial of the sensor systems used on autonomous cars, not because of what it does or how it works, but because it's one that Tesla doesn't currently use. Tesla is always good at getting people's attention, so their habit of ignoring lidar gets some notice.

It is a little odd that they'd shun lidar, because lidar is an incredibly powerful tool that can help a moving machine sense the world around it. Lidar stands for Light Detection and Ranging, and can be thought of as a sort of light-based radar. Lidar uses low-intensity, non-harmful, and invisible (to our meaty eyes) laser beams, which are pulsed at a target (or, in the case of most autonomous cars, all around it, in a full 360-degree sphere) and the reflected pulses are measured for return time and wavelength to compute the distance of the object from the sender.

In practice, lidar can produce some very detailed, high-resolution visualizations of the environment around a self-driving car. Here is an example of a visualization[30] of lidar data:

LUMINAR

Impressive, right? The lidar is often detailed enough to make out different surface textures and fairly small details on passing cars, and even things like potholes and manhole covers in the road.

Lidar units are also the most likely things to really challenge designers of future autonomous vehicles, since they require a high vantage point and an unobstructed 360-degree view. That's why lidar units are most commonly seen as domed objects mounted on roof racks on autonomous test vehicles. They're not exactly sleek, but they provide very good information about the world around them, and they're a good bit less affected by ambient light or weather conditions than camera-based vision systems. In fact, lidar could detect, say, a black-clad motorcyclist on a black motorcycle in the dark—you know, like Batman on the Batcycle—far better than a camera setup could.

There's another big advantage to lidar: while camera systems, even binocular camera systems, require computer time and algorithms to translate the array of pixels from the camera into a three-dimensional spatial map, the information detected by lidar is inherently three-dimensional data. That means that a lidar image of a car's surroundings can require significantly less processing time than a camera image, which translates to a faster response time for the autonomous vehicle.

Lidar is still a relatively new technology, and as such it's still not cheap; preparing lidar systems to withstand the brutal life of a typical automotive component is still in development.

GPS

GPS is pretty familiar technology to most of us by now, since it's the reason we all get lost so much less often than in decades past. The Global Positioning System relies on a constellation of satellites circling the earth to let us know exactly where we are at all times. Autonomous cars will use GPS heavily not just to navigate, but also to drive, since GPS lets the car can know what the road is about to do before it even gets there.

GPS allows autonomous cars to plan ahead, to be ready to slow down for a hairpin turn or speed up to jump an opening drawbridge. Just kidding. I mean, it'll be able to know where those drawbridges are, but I suspect you'd have to do some serious hacking to convince your autonomous car just how boss it would be to jump the bridge like a cop car in a clichéd '70s buddy-cop movie.

Standard GPS alone is only accurate to within one to three meters, or between about three and ten feet. If you, as a driver, were only able to accurately steer within three to nine feet, there's no way in hell you'd still have a license. Many, many things can fit within three feet of a car, including human beings, dogs, bicycles, oil drums full of acid, or chili, or both, a bear, and so on. That's just not good enough.

So, to compensate for GPS's relatively coarse resolution (I mean, considering that it's pinpointing a location on the entire planet, three to nine feet is pretty damn good; it's just not good enough for driving) autonomous car designers have come up with something called "localization."[31] There's a number of ways to accomplish localization, but most methods rely on the car's other sensor systems, with the goal of getting the car's location pinpointed to within ten centimeters or so.

Some methods use a technique called "particle filters," which seed a given map with some number of virtual particles at known locations. The particles can be thought of as possible locations of the vehicle. As the vehicle moves, sensors provide data about the speed and angle at which the car is moving, and the associated cloud of particles moves along with it, with the particles' locations compared against known landmarks on the map that are measured via the car's lidar systems.

Other methods can use the car's camera systems and exciting things like an "algorithm based on the probabilistic noise model of RSM features,"[32] and other gleefully geeky stuff like that. It's not a trivial problem, but there are many viable solutions.

GPS is also the technology that autonomous vehicles will use to report their locations back to any number of possible organizations: local law enforcement, your insurance provider, the carmaker who built the car, the company that developed the driving software for the car, and any number of other companies, advertisers, and market research groups that any of those organizations in the chain could have sold access to your car's location information to.

Autonomous cars will likely mean that no trip you take in them is ever going to be completely private, and we shouldn't even kid ourselves into thinking otherwise. Everyone will be watching everything, everywhere.

COMMUNICATION AND COMBINING EVERYTHING

All of these different forms of world sensing are combined to create an overall, composite image of the surrounding reality for the car. Some call this "sensor fusion,"[33] and it's especially important because each method, individually, has some pretty significant flaws and limitations that could cause real problems in practice.

Supplementing the sensor data is communication, both between cars on the road and from more centralized sources. Individual cars will communicate with others in their vicinity, a process known as vehicle-to-vehicle (or V2V) communication. There's already bandwidth set aside on the radio spectrum (at least in the United States) to accommodate this data traffic, the 5.9 GHz band. Europe has settled on the same basic frequency, though Japan is using 5.770 to 5.850 and 715 to 725 MHz.[34]

The thinking is that vehicles close to one another should share some essential information about what they're doing: such as speed, destination, predicted path, and any information regarding road or safety issues. Doing so will allow the cars to work together to find the optimal, most efficient traffic pattern and be hyper-aware of what everything around them is doing and/or planning to do at any moment.

Infrastructure-based information could be communicated as well, informing cars of the status of such things as traffic signals, traffic density, and lane closures. Emergency vehicles would be able to alert cars stuck in traffic to their presence, allowing cars to clear a lane in an orderly manner for vehicles like ambulances, fire trucks, or pizza delivery vehicles in cases where more than three toppings are specified, for example.

Really, the more information available to cars on the road, the better, and communication will allow autonomous vehicles to act as parts of a self-modifying system to maintain optimal traffic flow and, ideally, eliminate many of the traffic issues that so annoy us today.

The Car Connectivity Consortium and other groups of major car manufacturers are working together on this, which is good, since eventually a global standard should be established. It's also possible that a standard communication protocol could be used for human-driven cars as well, even possibly including systems that could be retrofitted to older cars. Even the primitive and crude death traps that I personally prefer to drive could, theoretically, be outfitted with a unit that sends such information as speed, throttle position, steering angle, and GPS location, all of which would help surrounding autonomous vehicles prepare themselves for the presence of the loon whipping a sixty-year-old Volkswagen around the streets.

SELF-DRIVING CARS DON'T REALLY GET TO BE DIRTY

One often overlooked detail when people imagine a future popu-lated with autonomous cars is that for them to really work effec-tively, we won't be able to let them get all that dirty, at least not in certain spots. Human-driven cars can withstand being pretty damn filthy, as long as the windshield stays mostly transparent, and lights are relatively free of filth, at least at night.

Autonomous cars are not likely to be so forgiving. Cameras are easily fooled or confused by the presence of dirt or leaves or other objects in their view, and a somewhat dirty windshield that most humans could look through and understand what they see just fine could absolutely confound an artificial vision camera system.

The same goes for radar emitter windows, lidar domes, and any other sensors on the car that require the ability to send and receive signals in the electromagnetic spectrum. Weather and the normal accumulated grime of life on the open road can shut down a car's ability to pilot itself quite quickly.

As of today, there's no one good answer for what to do about this, other than to try to keep things clean. That doesn't help if, say, a camera window gets splashed with mud or a lidar dome suffers an attack from a pigeon with severe digestive problems. Since these filthifying events can happen while a car is in motion, perhaps even at a high rate of speed, autonomous cars will likely need systems in place to clean and/or protect their sensors.

We already do some of this for human-driven cars: windshield wipers and washer systems are usually good enough to deal with weather and a bit of dirt, and some cars have similar wiper/washer systems for headlamps. Autonomous vehicles will likely utilize similar systems, like washer nozzles for radar or camera windows, or a sort of wiper "ring" for lidar domes, or . . . something.

Sensor cleanliness is one of those things that may seem trivial at first, but has the potential to play a key role in a life-or-death situation. Sensor cleaning systems will need to be able to react automatically and quickly, especially in cases where multiple systems are affected simultaneously, not at all an unlikely proposition considering how an entire car can be affected by weather or accidents like splashing through a muddy puddle.

It's like if you were driving in the mud during a rainstorm in an open car: it'd be hard to see, and it'd be highly likely that something would splash in your face to make it even harder. It's likely that any safety standards for autonomous cars will include a set of standards for sensor cleaning and/or protecting systems.

Really, autonomous vehicles will need to be self-aware. This isn't the sort of self-awareness that leads to your car one day refusing to drive until you answer the question displayed on its dash that reads, "Tell Me What This Thing Called 'Love' Is," that science fiction loves to wonder about, but it is a sort of self-awareness nevertheless, and as such will be pretty amazing.

Autonomous cars drive in essentially the same way we do: they look around, as carefully as they can, and they move, hopefully in the right direction and hopefully without hitting anyone or anything. These actions show how incredible it is that we humans can drive as well as we do, even without doing a lot of complicated math in our heads.

Human ability to react at high speeds, gauge stopping distances, sense vehicle weight shifts, and, you know, just *drive*, all while singing along to a Styx album at the top of our lungs is pretty damn impressive. It's no wonder that it's such a complex problem to solve for autonomous cars, and it's no wonder we're not quite there, yet.

We've come incredibly far, but there's still an awful long way to go.

CHAPTER 4

SEMIAUTONOMY IS STUPID

AS I ASSUME YOU GATHERED FROM THE TITLE OF THIS CHAP-
ter, I have a pretty strong opinion on semiautonomous driving systems in cars, such as the currently available semiautonomous systems from companies like Tesla, Volvo, and GM. I suppose there could be one additional level added to the list, albeit a very hypothetical one:

Level 6: Full Automation and Determination

The one thing no autonomous car is planned to be able to do is to actually decide *where* (and, I guess, *why*) it should drive. Even a Level 5 autonomous car will just sit in a parking spot, doing nothing, until it's given a destination, either by a human or some human-produced algorithm or schedule.

The absolutely ultimate self-driving car would decide where it wants to drive and just do it, not only not requiring any human intervention while driving, but not needing any human input, ever. Such a car would not have to care about people one bit, and would have no need to serve us at all. It would be a Level 6 car, and if Level 6 autonomous cars become available in any capacity, it's probably safe to assume that either things have gone very bad for humanity, or the world portrayed in the Pixar film *Cars* series has come to pass. Either way, humanity is boned.

———

Okay, since we now all know exactly what all the levels of autonomy are, I can get back to trying to convince you that the current cars sold as semiautonomous—that is, with Level 2 autonomy— are at best stupid and at worst downright dangerous.

Semiautonomous systems that require a driver to be focused on the task of driving, without actually or fully driving, are strange things, really. I've driven cars with these systems on a number of occasions, and it has usually been my experience that the act of focusing and paying attention to not-driving while being ready, at any moment, to drive, takes far more effort and attention than just, you know, driving. I mean, if your hands are on the wheel and your attention is focused on the road, you really may as well be actually driving. Driving comes with years of experience and muscle-memory to make it an almost automatic, lower-brain activity; I've found hovering over a car's controls, trying to second-guess what it's about to do just in case you need to take control, to be more stressful than just driving the car. When you're driving normally, you generally trust yourself enough to relax and drive. When you're nannying a Tesla on Autopilot or a Volvo using Pilot Assist, you're not really ever relaxed, because you're placed in a murky gray area between being in control and giving up control, and that's an awkward place to be.

Now, I don't mean to be entirely dismissive here; if you think of these semiautonomous systems less like a system that is capable of driving a car with a lot of supervision and more like a very comprehensive set of driver's aids that, really, just let you be a much worse driver, then these systems start to make more sense.

I was driving Volvo's system on a Swedish highway a while back, and decided to try this very approach. In the context in which a human driver is driving, but somewhat sloppily and inattentively, the system does a pretty good job of cleaning up around the edges and making things safer. You can drive and pay more attention to the surrounding, non-road landscape than you normally would. You can work a little less hard and focus a little less intently, and with the Level 2 assistance, still be relatively safe.

Now, I can't really say I recommend this, just that it's a way of thinking about Level 2 autonomy that makes a little more sense than nervously watching what the car is doing at every moment. It's sort of like Level 2 is a compensator for poor or perhaps even slightly buzzed driving, and, of course, none of those are good things. If used on a very regular basis, it's not hard to imagine that an eventual erosion of driver skill could be a result, since the demands on the driver are significantly reduced when the system is operating at its full potential.

It's when Level 2 systems seem to be working at their best that the most potential for trouble exists, because that's when the systems can deceptively seem fully autonomous, and that's when people begin to trust something that's not capable of living up to that level of responsibility. This perception doesn't just come from dazzled drivers; many of the companies that sell cars with Level 2 autonomous systems cast their capabilities in a light that suggests much more advanced abilities than is warranted. Look at Tesla's website,[35] for example:

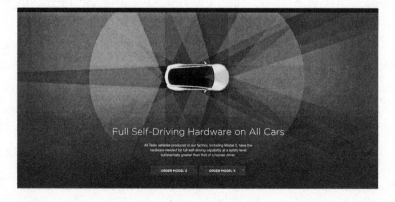

"Full Self-Driving Hardware on All Cars" reads the headline on their Autopilot page, and that sure as hell sounds like they're talking about something that can really drive itself, fully. "Full Self-Driving" is a powerful phrase, and is one that I think very much misrepresents the capabilities of their Level 2 Autopilot system.

Tesla is by no means alone here; a study by Thatcham Research[36] in October 2018 of 1,567 car owners from China, France, Italy, Germany, Spain, the United States, and the United Kingdom found that 71 percent of motorists believe that they could just go out and buy a fully autonomous, self-driving car immediately if they so chose. The truth is they couldn't then and can't now, because none existed for commercial sale, and that remains true.

Eleven percent of these same people also felt "tempted to have a brief nap" while using semiautonomous highway driving assist systems. This is, of course, a huge problem.

The consequences of putting too much trust in a system that has never been meant to handle the full responsibilities of driving can be fatal. There have been at least three[37] fatal accidents involving a Tesla in Autopilot mode, and a number of other nonfatal accidents. The theme among all of these wrecks is that the drivers put too much faith in the Autopilot system, overestimating its capabilities and, as a result, did not pay sufficient attention to what the car was actually doing.

This doesn't mean the drivers were fools, by any stretch, or even unfamiliar with technology. The fatal wreck that occurred on March 23, 2018, involved a thirty-eight-year-old Apple engineer. This was not someone unfamiliar with the fundamental technology involved, and even though the driver had complained to his Tesla dealer about the behavior of Autopilot in his Model

X before,[38] at the time of his wreck Tesla claims[39] his hands were not detected on the steering wheel.

In this particular wreck, the Model X, on a clear day with excellent visibility, veered into a concrete median on a highway near Mountain View, California. There wasn't anything in particular about the driving route or environment to suggest that the Autopilot system may have had trouble, but it did, and quite dramatically. The same goes for an earlier fatal Autopilot wreck in May 2017, when a Tesla Model S on Autopilot somehow didn't see a tractor trailer, and the driver didn't seem to be paying attention, either.

Autopilot, like any Level 2 system, has limitations, often pretty significant ones. I don't think Tesla's Autopilot system is technically unsafe at all; I'm sure if used properly, it has the potential to help make driving much safer, overall. But I do think Autopilot is *conceptually* unsafe, as are all Level 2 semiautonomous systems.

The problem is that people just don't work the way a Level 2 autonomous system demands. The concept that you can have a system that, on a relatively high percentage of highway roads and situations, handles most of the work of driving—and yet expects the person in the driver's seat to stay ready and alert—is dangerously wishful thinking. The only thing that moves faster than a Tesla in Ludicrous mode, Tesla's bonkers-fast driving mode, is complacency, and a Tesla owner, smitten with the technological promise of the car and Tesla's own marketing (the name "Autopilot" hardly suggests the need for constant attention) is very likely to allow their mind and attention to wander, even if they are keeping a perfunctory hand resting on the steering wheel, giving the car a false sense of security that they're on the job, just like the car gives to the driver, forming a dangerous feedback loop of false confidence.

GM's Super Cruise marketing is a bit more cautious, perhaps having learned something from Tesla's experiences. On the Cadillac website's Super Cruise section,[40] the ability to drive hands-free is certainly shouted, typographically, but so is the system's demand for the driver's attention, which GM has made a feature of their system by installing a driver-facing camera and head-tracking software.

The good thing about GM's head-tracking system is that by touting it as a feature that they can market, GM can reinforce the idea that the system requires constant attention on the road, even if they do allow hands-free operation. It's also a lot harder to spoof or fool than systems that just sense the presence of hands on the steering wheel, which can sometimes be faked with weights tied to the wheel.

Just in case you don't think anyone would be dumb enough to try to defeat the hands-on-wheel safety requirement of Tesla's Autopilot, I'd like to let you know that at least one person has fooled his Tesla into thinking there were human hands on the steering wheel by jamming an orange into the wheel.[41] Do not underestimate the genius of stupidity.

Also, there's a more cynical reason GM may have chosen to use a camera-based face-tracking system for their semiautonomous tech: litigation. If the car has a visual record of exactly where you were looking at the time of a wreck, then it's a lot easier for GM to argue that the driver was at fault, because you could clearly see that they were looking down at their lapful of spilled chili or working on their ship in a bottle or whatever.

Level 2 autonomous systems also demand that the person not-quite-driving be ready to take over in an instant if the car gets into trouble. This trouble could be something the car's electronics are aware of, like an issue with a sensor or a change in environmental conditions that prevents the car's sensors from gathering enough information. In these cases, it's possible for the car to provide a warning and at least a tiny bit of notice to whomever is moistening the driver's seat, so they can be sure they're ready to take over control of the vehicle.

A situation like that may not always be the case, though. It's possible that a semiautonomous car may find itself making bad decisions without even realizing it, as in the case of the Tesla Model X that crashed into the highway barrier. The system didn't even realize anything was wrong until it was too late, and the (possibly) inattentive driver didn't realize, either. In either case, the driver needs to assess the situation instantly and take whatever driving actions are necessary to stay safe.

If we think about this realistically, this seems like, at best, a pretty iffy idea. Let's say you're using your Level 2 Autopilot or Pilot Assist or whatever system you have, and it has been working great for several hours of a long, boring highway trip. I think if you've been sitting behind the wheel of a car that's been effectively driving itself for three or so hours it's not unrealistic to assume that maybe, just maybe, your mind has been wandering, and, even if your hands are on the wheel and your eyes are facing forward, out the windshield, perhaps your mind isn't exactly engaged in the task of driving. One could hardly blame you, really, since you haven't really been driving at all, just sitting there, staring out a window at a monotonous landscape, a never-ending ribbon of asphalt stretching ahead of you into the horizon.

Would you really feel like you'd be in a position to spring into action like a superhero if the car started doing something that could end in a wreck? Maybe? But maybe not. That's a lot of gambling to be doing behind the wheel of a car going 80 mph.

Level 2 autonomy relies on qualities in people that we just

don't excel at: constant attention to something you're not really entirely controlling, and being able to instantly take stock of a very fast-paced, high-pressure situation and make the right decisions—along with the associated physical motions—despite whatever adrenaline and panic may have been shot into your brain by the panic of it all. I wouldn't trust myself to handle either of these with any real success, and yet that's what these semiautonomous systems demand.

So, if Level 2 autonomy is incompatible with some pretty fundamental human characteristics and abilities, then what options do we have? Should Level 2 systems be regulated off the market, and consumers just wait until companies develop at least Level 3 autonomy, with its ability to safely handle stopping in situations when the machine determines it's unable to drive?

I'm not really sure if that's the way to go, even though I do think Level 2 systems are inherently flawed. I think it may be possible to develop a sort of hybrid Level 2 system that isn't dependent on the vagaries of the dumbass behind the wheel in situations when the autonomous system isn't up to the job. In fact, the use of the word "job" here I think is key; the only way to even remotely ensure that a human driver is paying attention and absolutely ready to take over as needed is by making that task into a job.

Sure, you could go old school and have that job filled by a personal, hired chauffeur, but if that's the case, you may as well just pay them to actually drive, and then you can get something much older and cooler than a Tesla, like a lovely chauffeur-driven 1973 Chevy Vega. No, I think there's a potential new job category for on-call/on-demand human drivers, ready to take over at a moment's notice, and they don't even need to be anywhere near the car.

The answer may be remotely driven vehicles. Remotely driven armed drones fly to the Middle East from the middle of America's fried-food heartland in Kansas on a routine basis, so why not employ the same basic technology for something less, you know, warlike?

Car companies have been experimenting with the idea of remotely driven cars for years,[42] and it has been proven possible.[43]

Even with sensors obscured by weather or debris, a human driver, using the lane-departure camera system or some other similar windshield-peering camera, would likely be able to drive, and a human's greater driving experience and the nature of the brain and visual system means that conditions that might confuse a computer would be no big deal for a human.

We've all driven in bad visibility conditions in lousy weather. It's no picnic, but humans can certainly do it. Overall, a computer-based autonomous system will definitely have better reflexes and drive more safely and rationally than a human could, but there are still situations where a pair of moist human eyes—even if they're thousands of miles away—can't be beat.

A company like Tesla would have a facility with some number of specially trained remote drivers in front of what would essentially be fancy driving-simulation rigs. When an autonomous vehicle determines it cannot continue, it would contact the remote driving facility, and the car's controls would be patched to a remote driver's terminal.

This method would be far more reliable than expecting the person in the car to be ready to suddenly take over, because it would make the role of the vigilant non-active-but-ready-to-go driver into an actual job, with responsibilities and supervision and an actual chance in hell of being ready to take over should a situation demand it.

The remote driver would announce to the occupants of the car what's going on, and would make sure that the occupants confirm verbally that someone is awake and available to take control if there's a communications issue. Then, the remote driver would pilot the car to its GPS-set destination, or until the autonomous system determined it was able to resume control. The remote driver's actions would be logged and incorporated into the autonomous system's machine-learning systems, in the hopes of making the autonomous system even better.

A decent-sized call center could likely provide on-call support to a large number of semiautonomous cars, although I admit I haven't done the math, since pretty much every number would be speculative. There's also something to be said for the possibility of job creation, which would help counter some of the fears that increased autonomy in cars would lead to job loss.

Really, though, emergency remote driving of a semiautonomous car is just a Band-Aid. The real problem is, as it always will be, that Level 2 semiautonomous systems are inherently flawed. They reside in that amorphous zone between competence and disaster, being both capable enough to be useful in many situations and as a result, inspiring a confidence that is far too generous for what the system's true capabilities are.

If you were to assess the pros and cons of Level 2 semi-autonomous driving systems, I don't think the result would be that flattering to the systems, even while we acknowledge just how technically impressive most of these systems are. Sure, they can take over a lot of the work of the most tedious, boring sort of driving, but they replace the driving tasks—ones that most of us have been doing for years and can now perform with such ease and confidence that the process can be relaxing—with a new set of tasks involving monitoring a partial autonomous driving system that the car's owner is not likely to fully understand.

The downsides are pretty dramatically nontrivial, too. The systems are not infallible, and are deployed on cars that do not have a full suite of autonomous sensors and equipment. No Level 2 car sold today, for example, uses a lidar system, and so far Tesla has stated they have no intention of using such a system, even in the future.[44] They're not fully equipped for the demands of autonomy.

The Level 2 cars available today offer an interesting and dangerously tantalizing taste of a possible autonomous future, and that's both their biggest advantage and disadvantage. Yes, their deployment on public roads provides a lot of data that will undoubtedly help the development of future, more advanced autonomous vehicles, but even so there's a lot of debate about whether deploying these cars in the public realm unwittingly makes drivers on public roads participants in a vast beta test-ing program.[45]

All of this is exacerbated by the way in which cars with systems like Tesla's Autopilot are marketed and the way they're understood in popular culture: in both cases, as far more capable systems than they actually are.

There are no fully autonomous vehicles currently available for purchase—arguably, some prototype test vehicles exist that could be considered fully autonomous, at least in some controlled circumstances. Sure, if you're reading this in the ruins of an old bookstore in the year 2876 as you hide from the blood-craving completely autonomous Teslas that roam the streets, hunting, then this fact may strike you as odd. But it's true, and it's worth remembering, especially the next time a tipsy Tesla zealot corners you at a friend's dinner party. At this point in time, Level 2 semi-autonomous cars can do impressive things, but they're not there yet, and that means we can't treat them like they are.

CHAPTER 5

THEY'RE ROBOTS, NOT CARS

PERHAPS THE BIGGEST REASON WHY THERE ARE SO MANY questions and concerns about the coming of autonomous cars is because we all seem to insist on referring to them as "cars." Now, I'm not saying that they're entirely *not* cars, because, of course they are. They're evolutions of cars, and the ones we have today are based on conventional cars, for the most part. But an autonomous car is, really, no longer a car. It's a robot. It's a robot whose primary job is to move cargo, whether that cargo is boxes of rubber vomit to a rubber vomit sales and distribution company, or that cargo is a punk band of scrappy teens with some real talent, if they'd only realize that their lead singer is a preening douche who's holding them back.

To the robot that we call an autonomous car, it doesn't really matter. As far as the robot knows, it's all just cargo, whether it's people or stuff, and that cargo has to get to a location quickly and safely.

I'm really not crazy about the term "autonomous car," now that I think about it, because it clearly defines what we're talking about as a "car," which I'm not sure makes sense. Autonomous cars will not really be cars, and will be fundamentally different from automobiles as we know and love them today. As I mentioned earlier, at possibly painful length, a car is really a prosthetic extension of our human bodies. It's a physical thing we control with our bodies.

An autonomous car is decidedly not that. We will all ride

in autonomous cars, not drive them. Fully autonomous vehicles won't ask anything of us other than to provide an end destination and not do anything crazy like jump out of the car while it's in motion. In this sense, they differ pretty substantially from cars as we currently know them. They are assigned a task by a human (getting to a destination) and then seek to achieve that goal via sensors that communicate information about the outside world and a set of preprogrammed algorithms and method that allow them to adapt to their environment and circumstances as best they can. Autonomous cars are probably best classified as robots. In fact, in my opinion, we should rework our vehicular classification system to accurately reflect what a robot car is.

Okay, maybe that won't be quite as much fun as I imagine, but you get the idea: the difference between whether a vehicle is driven—as in fully controlled, not just told what destination is desired—or not, is a fundamental factor in determining what that vehicle is.

That's why I think we need to train ourselves to think of autonomous cars not as a type of car, but as a type of robot, a transportation robot. In fact, I think for the rest of this book I'm going to call them "robotic vehicles," because that frees them from the constraint and baggage of the word "car," which colors our thinking when it comes to how we deal with autonomous cars. Sorry, robotic vehicles.

It's a shame that robotic vehicles' initials are RV, because everyone already associates that with recreational vehicles. Dammit, that's a problem. Maybe we can use "robotic transportation machines?" RTM? That's sometimes used to mean "read the manual," but usually you see that as RTFM, where the "F" means exactly what you know it means. So maybe let's try that.

So, with the terminology mostly out of the way, let's consider

RTMs in a larger context, freer from the constraints of how we think about cars. If the future of RTMs goes as the most optimistic plans suggest, then RTMs will be the first real mass-deployment of sophisticated, self-acting robots in human history. Sure, the world is filled with other robots, but by far the vast majority of these spend their robo-lives in factories and industrial facilities, doing work and interacting with human workers, but these are still specialized machines not in mainstream society.

And while there's been consumer robots in people's homes for a while now, mostly in the form of vacuums and other floor-cleaning robots that look like the oversized, scurrying offspring of a hockey puck and a turtle, there's one huge difference between the small, simple robots we currently deploy in people's homes and what could very well exist one day: a transportation robot the size of a car can kill you.

So, perhaps it's better to look at the upcoming age of robotic transportation machines as the first mass deployment of potentially lethal autonomous machines in human history. And by "better" I mean "alarmist."

Robotic cars, by virtue of having the size and strength of a car, have the potential to be dangerous, just like any car driven by a human has the potential to be dangerous. The difference is that with conventional, human-driven cars, we take a lot of time and effort to train the organic computers that drive the cars to understand that driving like an idiot or with the intent to cause harm is very, very bad. We do this through culture, religion, education, and our legal system, and we've been teaching these same fundamental concepts since long before cars were even a thing. This is thousands and thousands of years, since the fundamental ideas of not doing stupid or evil things to kill people are foundations of most major religions.

But for machines, we'll need to codify these ideas. Perhaps the most famous codification of ethical rules for machines that have some degree of autonomy are the Three Laws of Robotics from the famed science fiction writer Isaac Asimov:[46]

1. A robot may not injure a human being or, through inaction, allow a human being to come to harm.

2. A robot must obey the orders given it by human beings except where such orders would conflict with the First Law.

3. A robot must protect its own existence as long as such protection does not conflict with the First or Second Law.

A bit later, Asimov added a zeroth law that outranked the other three:

0. A robot may not harm humanity, or, by inaction, allow humanity to come to harm.

We'll soon discuss these laws and their implied ethics more thoroughly, but it's worth noting that a few years after Asimov had written *I, Robot*—where the Laws of Robotics were first introduced—he wrote a short story called *Sally*[47] that ran in the May–June 1953 issue of the pulp sci-fi magazine *Fantastic*.

Sally is a story about autonomous cars, and proves to be remarkably, maybe even shockingly, prescient in many ways. Some of the ways are simple, like getting the general timeline close: the story takes place in 2057, and references are made to the oldest automatic (the story's term for autonomous cars) being from 2015. If we count the Level 2 semiautonomy available in several cars from 2015, that's a pretty damn good guess.

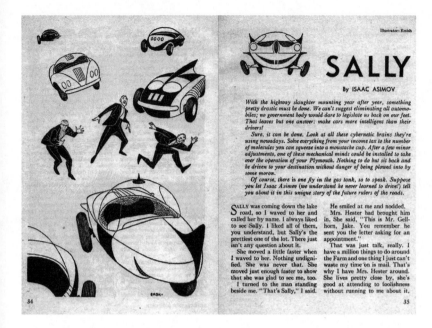

Other predictions are more involved, more detailed, and
while they're predicting events that haven't happened yet in reality, they reflect a lot of what the conventional thinking about the
future of autonomous cars contains. The story is about a sort of
retirement ranch for robotic cars that border on the sentient, so
it's not *completely* prescient, but there are a few key points I think
Asimov gets right. We can find most of them in these paragraphs
near the beginning of the story:

> The thought makes me feel old. I can remember when there
> wasn't an automobile in the world with brains enough to find
> its own way home. I chauffeured dead lumps of machines
> that needed a man's hand at their controls every minute.
> Every year machines like that used to kill tens of thousands
> of people.

The automatics fixed that. A positronic brain can react much faster than a human one, of course, and it paid people to keep hands off the controls. You got in, punched your destination and let it go its own way. . . .

Of course, the automatics were ten to a hundred times as expensive as the hand-driven ones, and there weren't many that could afford a private vehicle. The industry specialized in turning out omnibus-automatics. You could always call a company and have one stop at your door in a matter of minutes and take you where you wanted to go. Usually, you had to drive with others who were going your way, but what's wrong with that?[48]

Here Asimov accurately describes the very simple operation of an autonomous car (punch in a destination, let it go), but also acknowledges that such cars will likely be much more expensive than conventional cars, and suggests that ownership of private autonomous cars will be limited to the rich, with less wealthy folk using the "omnibus-automatics," where you "call a company and have one stop at your door in a matter of minutes and take you where you wanted to go." That sounds a hell of a lot like Uber, which, of course, is conducting a lot of research into autonomous vehicles.

The emphasis on safety as the primary justification for autonomous car development is something we see a lot of today. Even though there are many reasons why people may desire an autonomous car—mobility freedom for disabled people or people who can't drive for other reasons: the ability to do work or other things while in transit, the possibility of driving home without worry while drunk, or simple human laziness—the safety angle is almost always the one most aggressively promoted because, let's face it, who's going to argue with the possible outcome of fewer people dying?

Asimov's story also suggests that there will be legislation to effectively eliminate human-driven cars from the road, which is a prediction I really don't want to see happen, for reasons I'll be getting into in more detail later.

Let's look at Asimov's prediction about the cost of robotic vehicles, because he's definitely onto something there. I can't even guess at how to add up the costs of all of the hardware and software of the extra sensor systems and the need for continual updates, improvements, and possible regulatory fees or other costs, and there's no way a robotic vehicle is going to sell anywhere close to the price of a human-driven car; it's going to far surpass it.

The truth is that replicating a human's ability to drive takes a lot of expensive equipment, and the cost of creating the traditional biological alternative is far less, and, ideally, is more fun. This means yes, robotic vehicles will be quite expensive, at least at first, until continued technological developments and economies of scale eventually bring the cost down.

But, until then, robotic vehicles will be for rich people, at least private vehicles. This suggests some interesting outcomes. If robotic vehicle ownership skews significantly toward wealthy people, then we may see a situation where the skill of driving becomes less common the higher the person's income. Wealthy people who have yet to learn how to drive could possibly avoid the step of learning how to drive altogether and go right into self-driving vehicles for their personal and private transportation.

Will the inability to drive begin to have some sort of status? Will people lie that they don't know how to drive to convey the illusion that they're rolling in it? It's possible; cars have always been very potent signifiers of status, and there's no reason to believe that will change any time soon.

Really, we've seen this situation before, in the early part of the twentieth century, when automobile ownership was primarily for the rich, and they would often employ a chauffeur to drive them around. The combination of owning a car that you have no idea how to drive, yet always being able to count on that car getting you to wherever you need to go has always been a surefire way to let everyone know how rich you are.

The idea that, for most people, their experience with robotic vehicles will be via shared cars has long been assumed by proponents of an autonomous future. There are lots of really good reasons to think that all cars should be shared, or even communal, and private car ownership makes little sense: private vehicles spend most of their time idle, so why not have a system in which the cars aren't almost always dormant? Fewer cars would be needed overall, which means fewer parking spaces needed, which means more room in cities, which means less traffic, and on and on. For these same reasons we could argue for the sharing of robotic vehicles, but I'm not totally convinced *all* cars should be shared, even if they can drive themselves.

Most of the benefits touted by champions of autonomous cars tend to be very quantifiable things: millions of fewer deaths because of driver-error–caused crashes, multiple suns' worth of energy saved from the optimization of trips, hectares of parkland opened up because we won't need parking lots—since we'll all be riding around in shared autonomous cars— and loads more productive time since we'll be free of the task of driving. It doesn't really matter that most of these figures have been pulled *ex recto* since nobody has any supportive data; the point is that it is hard data, or at least hard-looking data. Numbers, figures, that kind of thing.

While I'm sure many of these benefits may prove to be accurate at some point in the future when self-driving cars (wait, I need to remember to call these RTMs) are common, I don't think thinking about the advantages of RTMs over human-driven cars in these sorts of quantifiable, numerical terms even comes close to describing how the relationship between people and their cars actually works in the real world.

For public transportation, sure, numbers and statistics tell the essential story. But when it comes to personal transportation—and by that I'm mostly talking about cars—numbers and data are woefully inadequate. This is because automobiles have never, ever been entirely rational things. Considering how significant a purchase they are for most people, and how crucial a role they play in most people's daily lives, the rationale most people use to pick their cars is absurdly irrational, and very likely represents the greatest ratio of irrational decision-making of a large purchase that people will make in their lives, often multiple times.

If cars were absolutely rational, they'd be highly standardized, with as many interchangeable parts as possible across the entire range of vehicles. They'd have engines that would generate about 100 horsepower at most, because that level of power is completely adequate for almost all normal use. They'd be comfortable enough, and you'd see about as much styling variation as you find in your average selection of dishwashers or refrigerators.

There'd be some variety in terms of the level of equipment, comfort, and body type variations for different uses, but for the most part, all cars would be basically the same. They might even get sold along with houses in the same way that washers and dryers are now, and we're fine with that, because, deep down, nobody really gives a damn about a washer or dryer as long as they wash or dry.

This, of course, is not at all how cars are bought, sold, or owned in the real world—planet Earth, the birthplace of the car.

Cars are ridiculous things—which is exactly why I and so many other people love them so much and so irrationally. Proponents of RTMs (remember, that's what I'm calling self-driving cars) who think that the whole story—about how many potential lives can be saved—can be told by speculative statistics are, let's face it, fools. That's not to say saving many, many lives isn't important—of course it is, and always will be—but how those lives get lived is important as well, and much harder to quantify, but this will also be absolutely crucial in how RTMs get accepted and used because their target market will still be human beings.

Look at the sorts of cars you can buy today: Dodge sells a car called the Demon that makes 840 horsepower, can go from a dead stop to a mile a minute in 2.3 seconds, and doesn't really handle all that well. It's a machine designed to peel out of Dairy Queens and win at drag races. Rationally, it's absurd. Yet, with some significant compromises in comfort and economy, you can drive it to work. So go ahead and tell your spouse you need it.

Most people are not remotely qualified to handle a vehicle with that much power. There's no road in America where you can lawfully use it at anything close to its full potential. Unless you get one to seriously drag race, everything about that vehicle is pretty much wasted. But people still buy them, or its 707-horsepower sibling, the Dodge Hellcat, or any number of the other five-hundred-plus horsepower performance cars on the market.

People do, of course, use some very rational metrics when deciding to buy a car, but even then, cars are so laden with cultural associations that nothing is ever as easy as what's on paper. A Toyota Prius, for example, can be objectively thought of as a highly rational car. Its shape and basic design are primarily dictated by aerodynamic concerns, even if the front end of the latest one looks like an alien fish that you just fed a lemon wedge to and then told that you'd like to start seeing other alien fish.

It's got a very efficient hybrid drivetrain and has decent interior room. It's generally quiet, comfortable, and reliable, and has adequate power, and will absolutely do the job of personal transportation. Yet it's also so closely associated with a particular category of people, usually urban, educated, left-leaning, and all of the associated stickers, sandals, and dietary demands that go with that, that many people who would otherwise be very well served by such a car will reject it because the image of the car is counter to their own personal image.

In many ways, cars are far closer to fashion than any other bit of life-machinery we buy and use on a daily basis. What we choose to wear telegraphs a vast amount about who we are, and who we want to be perceived to be. Cars are the same way, which is unique for something that we don't actually wear on our bodies.

You could argue that personal electronics are similar—people wish to convey status or personal information about themselves if they, say, have an Apple laptop—but there's not nearly the same level of choice or granularity to the number of possible options as there are in the vast world of cars.

The idea that the ideal future involves fleets of shared-use autonomous cars does not acknowledge the true nature of how humans interact with their personal transportation, nor does it acknowledge the motives of carmakers. Shared RTMs make a lot of sense in dense urban areas, places where a high proportion of the population may not already own a car. This is a significant market, and I'm pretty sure there will, one day, be self-driving fleet vehicles to serve the transportation needs of the people there.

The idea that RTMs will bring about the death of private car ownership is taken almost as a given by many, including many very Serious Business media outlets like the *Wall Street Journal*,[49] which has published articles like "The End of Car Ownership," or the *Economist*'s "Why driverless cars will mostly be shared, not owned,"[50] or *Business Insider*, which in 2017 cited a study that predicted that in fifteen years only 20 percent of Americans will own a car.[51]

These predictions are pretty out there, I think. Do any of these people have kids? Have they seen how much equipment, junk, and random stuff hauling around a human child generates? Have they thought about the logistics of moving all that stuff from car to car? No, of course they haven't. Half the reason for having a car is that it is a mobile base for you and all the stuff you don't know where else to put. Sure, in some dense cities, not having a car means a lot less hassle and an easier life. But everywhere? No. No way. Even in some dense but vast urban areas, like sprawling Los Angeles, where a lot of the population could be served with fleet cars, for logistical and cultural reasons private car ownership will still be desirable. Ever had an awful job and eaten lunch or taken a nap in your car because it was your only refuge? Do you keep jackets and umbrellas and other useful real-world things in your car? Do you like to listen to your own music in your car? Do you sometimes do multiple things in the course of a day that involves multiple packages or parcels or items and you don't want to have to haul them around everywhere on your back like some refugee?

Of course you do, for all these things, because a car isn't just a transportation device. It's also a space. A location. A car is one of the few things that can be a means to a location and an independent location itself. Very often when we're, say, talking to someone on the phone while driving (using our hands-free options of course, with our hands firmly gripping the wheel at 9 and 3) and the person asks where we are, an entirely acceptable answer is "in the car." No further elaboration is often needed. You're in the car, and the space of the car is sufficient description about your surroundings and what you're likely to be doing.

There's no reason why driverless cars can't continue this in this role, and, very likely, even expand on it significantly. This is something I'll be covering in more detail later, but the idea that the rise in self-driving vehicles will mean the death of the privately owned car is the sort of idea that only someone who has never owned a car would have.

The truth is that the reality will likely be a mix of shared and private cars, and the roads will be filled with a mix of human- and machine-driven cars. And in this possible upcoming reality, it will still make sense to stop ourselves from thinking of robotic vehicles as cars, and to accept that they're robots.

Once we free ourselves from the constraints of thinking about autonomous cars as "cars," many possibilities open up. These machines could be errand-running robots that have the potential to significantly improve our quality of life, once we realize that, since they're robots, we don't necessarily have to be anywhere near them to use them.

Think about how many boring, stupid, but necessary errands you have to run in a given week. If you have a vehicle that can drive itself, why bother to do those errands yourself? Once robotic vehicles that have been proven safe start to make it onto public roads in any sort of numbers, *and* if there are laws that allow the vehicles to operate completely unmanned, then we will probably start to see fascinating new business models emerge.

The same process would work for any number of retail businesses. Why would you ever go to a pharmacy to pick up a prescription in person again? You could send your errand-bot to pick up packages from the post office, or you could even use the vehicle to courier packages to any address within the car's user-set range. Small businesses could buy fleets of five to ten of these and be able to offer deliveries without hiring drivers (a mixed blessing, I know).

Once you start thinking about it, what's to stop you from sticking your dog in the robot and sending it off to the vet? Or, since we're already heading that way, putting your kid in there and sending him or her off to school? If you trust the car with your kid's life when you're riding in the robotic vehicle with him or her, why should it be any different if you're not there?

The real potential for autonomous vehicles is not tied to how much we'll be riding in them, it's tied to how much we'll be able to avoid getting in them. Sure, doing whatever you feel like while your car is stuck in traffic is a lot better than driving in stop-and-go traffic, inching along at 5 mph with your clutch leg aching and your mind screaming out of boredom, but what's even better than that is not being in traffic at all. Let your robot car creep along at a walking pace to get you a drum of cheese balls and a pack of underpants from Costco. You have better things to do. Or, if not better, at least more fun.

This line of thought leads to the conclusion that there will likely be a market for robotic vehicles never intended to hold passengers at all. There's no reason why a cheaper alternative to a full-size robotic vehicle couldn't be a cargo-only errand-bot. A few years back, I explored this idea in the context of speculating on what a future Apple-branded car would be like[52] and came up with this simple design:

A car like this—though I don't think it'll be an Apple-branded thing—would be incredibly useful. It wouldn't need to be as large, expensive, or complex as a vehicle designed to safely convey people around, and as such would be cheaper and more accessible to larger numbers of people. Think about all the safety and comfort equipment you wouldn't need: airbags, seat belts, seats, interior padding, instruments or displays, even windows. Just external safety equipment and all the required indicator lights and that sort of thing. That's a lot of hardware the car simply wouldn't need to have.

Even if you retained a conventional, human-driven car for your own personal transport needs, a cargo-only robotic vehicle could make your life substantially easier. In fact, as an adjunct to a human-driven car, this could offer a lot of possibilities.

Let's say you have one of these, and a vintage small roadster, like a Porsche 356 or an Austin-Healey Bugeye Sprite—something small and fun, and maybe a bit finicky. One of these cargo-only robotic vehicles could still let you drive to a place like Home Depot to get lumber or large equipment, and then take it home by loading it into the robotic cargo vehicle, and having it follow your old Bugeye Sprite home.

You could get the most out of your pleasure cars by having the robot follow you on trips, filled with tools in case you break down, or, in real emergencies, the robotic vehicle could actually tow your vintage car if you can't get it going again. You could even take a road trip with your robotic errand-bot following you, and sleep in it at night, like a little camper.

When we stop thinking of autonomous cars as cars, what we can allow them to be or do opens up dramatically. They're robots, and there's no reason we have to restrict their potential by thinking of them as something else.

SIZE COMPARISON

In some ways, I can see the deployment of unmanned cargo or service-only robotic vehicles happening before we see fully autonomous passenger vehicles, because, freed from the burden of protecting human life, such robotic vehicles would be ideal for long-term testing of autonomous driving hardware, software, and methods. These could make excellent intra-city delivery vehicles, the sorts of things sold in fleets to companies like UPS and FedEx. Utility companies could also use these to check out remote areas with problems, sending back video of the situation, or deploying other robots designed to effect repairs, either via teleoperation or autonomy. There could also be postal vehicles, meter-reading bots, police patrol and awareness vehicles, mapping vehicles, food delivery and/or vending trucks, and so many other ideas I've yet to think of.

Really, the most useful autonomous vehicle in your life could end up being the one you'll never ride in.

CHAPTER 6

ETHICS, BEHAVIOR, AND BEING BETTER THAN PEOPLE ARE

A LOT OF THE DISCUSSION ABOUT A FUTURE FULL OF ROBOTIC vehicles has been concerned with the concept of ethics, and how we want these machines to behave. These are, of course, extremely important questions and concerns, but let's be real here: we're all sort of being hypocrites whenever we wring our hands over how we expect robotic vehicles to behave in morally or ethically difficult situations, where real lives are at stake.

We're hypocrites because humanity is basically a collection of all kinds of often miserable jackasses who wouldn't know the best ethical solution to the trolley problem if it shoved its ethical and hypothetical tongue in their nostril, and just about all of those miserable jackasses have car keys.

Oh, and in case you're not familiar with it, I'll explain the trolley problem soon. The recent interest in autonomous vehicles has made this fifty-two-year-old thought experiment surprisingly popular, so, don't worry, before you fling this book to the ground in disgust, you'll know what the hell everyone's talking about.

In the previous chapter, we lightly touched on the idea of ethics for robotic vehicles, referencing Asimov's laws of robotics. While those laws are a good conceptual start, they're not necessarily detailed or specific enough to be effective guidelines for future robotic vehicles, and it's pretty clear there's still a great deal of thinking that needs to happen. Robotic vehicles will be, as we mentioned before, generally car-scaled and with car-levels of power, which are more than adequate to make people very dead and unhappy if things go wrong. If we're to have these things roaming the roads of the world en masse, it's probably worth taking a bit of time to come up with some rules for their behavior that everyone agrees on.

The level of anxiety many people have when they consider these ethical issues is interesting. There's a certain weird satisfaction many people get, some sort of Luddite pleasure that comes with imagining scenarios where the artificial intelligences guiding all these robotic vehicles decide that they're sick of being ordered around by humans and revolt against us, wearing our severed heads atop their grilles like macabre hood ornaments.

Personally, I'm sick of these jokes and I feel like people's alleged panic and mistrust of machines is at best misguided and at worst a lot of attention-hungry theater. That does not mean that there aren't real dangers and issues associated with populating our cities and highways with thirty-five-hundred-pound self-directed machines; it just means that we have to remember what these things are actually and realistically capable of.

They don't really have any "intelligence" as we understand it, artificial or not. They may be able to do remarkable things, and exhibit behaviors that sure as hell look like they're thinking, reasoning beings. But the truth is they're just machines, reacting to inputs, environmental or otherwise, based on a set of algorithms and parameters designed by normal and likely smelly human beings. They'll never decide that they're sick of ferrying us all over town to do stupid things and eat in overpriced restaurants or see friends' embarrassing one-man shows, because they simply lack the ability to have opinions on anything, and will never decide to revolt and murder us all because they are fundamentally incapable of the creativity required to generate such an idea, or any idea at all.

Robotic vehicles may one day commit some terrible acts of destruction, but the impetus will be human-caused, either because of genuine malicious intent or deeply unfortunate mistakes. The robot may be the tool by which bad things happen, but it will never be the mastermind. In this way they're the same as conventional human-driven cars. A sick or evil person can easily turn a car from a transportation machine into a tool of death and destruction, and we've seen this happen for well over a century, pretty much as long as cars have been readily available.

So, while I don't think we'll ever need to seriously worry about a robotic uprising by the cars themselves, there is plenty else to worry about. With any networked computer system, there are very real and significant fears about security and hacking, as well as the normal mundane complexity of modern human civilization that offers nearly infinite opportunities for things to go dramatically and exuberantly wrong. Let's look into this aspect first, and think about how future robotic cars will deal with a confusing world.

This means we should probably address the trolley problem first, since almost every discussion of autonomous car ethics will bring this up, and I've put it off as long as I could.

The trolley problem[53] was first "officially" stated by the British philosopher, ethicist, and hilarious-name-haver Philippa Foot in 1967. Foot's original description of the trolley problem reads like this:

> Edward is the driver of a trolley, whose brakes have just failed. On the track ahead of him are five people; the banks are so steep that they will not be able to get off the track in time. The track has a spur leading off to the right, and Edward can turn the trolley onto it. Unfortunately there is one person on the

right-hand track. Edward can turn the trolley, killing the one; or he can refrain from turning the trolley, killing the five.[54]

The only absolutely unequivocal takeaway from this is that it deeply sucks to be Edward, and whatever they're paying him is not enough. Aside from that, the core of the trolley problem is the question: is it worth the death of one person to save five? As you can imagine, this could be important to a self-driving car if it finds itself in a situation where, say, it has six occupants inside it, and a truck suddenly pulls out in front of it. If the car continues on its current path, all six passengers will die. If the car veers sharply to the left, it could get around the truck, but there's a child on a bicycle in the next lane, a wonderful, beautiful child who gets good grades and dearly loves her mother and dog, and may one day invent a shampoo that will help you lose weight.

Should the robotic vehicle choose to veer into the child, killing her but saving the six passengers in the car?

For what it's worth, most studies seem to suggest that people are in favor of killing fewer people to save more people.[55] Of course, there are all kinds of ridiculous ways to complicate this, with wildly hypothetical situations, like what if there were fifty geriatric and ugly Nazis on an autonomous bus and if you want to save them you must sacrifice Jeff Goldblum and a philanthropist who wants to provide free tacos and eternal kittens to the world. Would the sheer number of lives saved be the deciding factor here?

Philippa Foot herself further complicated the trolley problem with this variant:

George is on a footbridge over the trolley tracks. He knows trolleys, and can see that the one approaching the bridge is out of control. On the track back of the bridge there are five people; the banks are so steep that they will not be able to get off the track in time. George knows that the only way to stop an out-of-control trolley is to drop a very heavy weight into its path. But the only available, sufficiently heavy weight, is a fat man, also watching the trolley from the footbridge. George can shove the fat man onto the track in the path of the trolley, killing the fat man; or he can refrain from doing this, letting the five die.[56]

Okay, even if we just accept the ridiculous idea that anyone without the last name of "the Hutt" can be large enough to stop a speeding trolley with their massive body, this variation doesn't really offer all that much more insight. The key difference is how direct the killing action is; in the first case, Edward is not actually directly killing the man on the trolley tracks if he chooses to switch the trolley's tracks. He's just taking an action that will result in the man's death. George, on the other hand, is actually and directly murdering the poor fat man to save the five people. Is this difference actually significant? Does anything about the trolley problem really matter?

The truth is that, in reality, I don't think the trolley problem is really a likely dilemma that autonomous cars will face. Sure, they may end up in situations where sacrifice of life is unavoidable, but the idea that the robotic vehicles will have access to all the information that makes up the trolley problem—the number of passengers in the vehicle, specifically—is by no means assured, and as such is not likely to be a factor in the cars' decision making.

Plus, there are so many more variables, that a decision based solely on the number of people potentially harmed is extremely unlikely. In reality, a robotic car attempting to avoid a collision will be factoring in a vast number of variables, including the relative braking distances for each possible path, other obstacles that may be in the possible directional vectors that could help or hinder the outcome, as well as such factors as weather, visibility, and the ability of the car's structure to protect the occupants.

The trolley problem is too much of an abstract thought experiment to be really useful or worth worrying about. The real world is so much more messy and complicated. But that doesn't mean no rules can apply. Some set of ethical rules are absolutely necessary, to form a framework and a place to start decision-making when difficult situations inevitably arise. Ideally, these rules would be universal, so that the behavior of robotic vehicles is at least somewhat predictable. If we're to have large numbers of robotic vehicles around, they all need to be playing by the same set of rules.

So far, only one country has attempted to codify a set of rules. In June 2017, Germany's Federal Ministry of Transport and Digital Infrastructure issued a report from their Ethics Commission on Automated and Connected Driving. This report codified twenty "ethical rules for automated and connected vehicular traffic." Here are the rules they came up with, in full:

1. The primary purpose of partly and fully automated transport systems is to improve safety for all road users. Another purpose is to increase mobility opportunities and to make further benefits possible. Technological development obeys the principle of personal autonomy, which means that individuals enjoy freedom of action for which they themselves are responsible.

2. The protection of individuals takes precedence over
    all other utilitarian considerations. The objective is to
    reduce the level of harm until it is completely prevented.
    The licensing of automated systems is not justifiable
    unless it promises to produce at least a diminution in
    harm compared with human driving, in other words a
    positive balance of risks.

3. The public sector is responsible for guaranteeing
    the safety of the automated and connected systems
    introduced and licensed in the public street environment.
    Driving systems thus need official licensing and
    monitoring. The guiding principle is the avoidance
    of accidents, although technologically unavoidable
    residual risks do not militate against the introduction
    of automated driving if the balance of risks is
    fundamentally positive.

4. The personal responsibility of individuals for taking
    decisions is an expression of a society centred on
    individual human beings, with their entitlement to
    personal development and their need for protection. The
    purpose of all governmental and political regulatory
    decisions is thus to promote the free development and
    the protection of individuals. In a free society, the way in
    which technology is statutorily fleshed out is such that a
    balance is struck between maximum personal freedom
    of choice in a general regime of development and the
    freedom of others and their safety.

5. Automated and connected technology should prevent
    accidents wherever this is practically possible. Based on
    the state of the art, the technology must be designed in
    such a way that critical situations do not arise in the
    first place. These include dilemma situations, in other
    words a situation in which an automated vehicle has to
    "decide" which of two evils, between which there can
    be no trade-off, it necessarily has to perform. In this
    context, the entire spectrum of technological options—
    for instance from limiting the scope of application to
    controllable traffic environments, vehicle sensors and
    braking performance, signals for persons at risk, right
    up to preventing hazards by means of "intelligent"
    road infrastructure—should be used and continuously
    evolved. The significant enhancement of road safety is
    the objective of development and regulation, starting
    with the design and programming of the vehicles such
    that they drive in a defensive and anticipatory manner,
    posing as little risk as possible to vulnerable road users.

6. The introduction of more highly automated driving
    systems, especially with the option of automated
    collision prevention, may be socially and ethically
    mandated if it can unlock existing potential for damage
    limitation. Conversely, a statutorily imposed obligation
    to use fully automated transport systems or the causation
    of practical inescapability is ethically questionable
    if it entails submission to technological imperatives
    (prohibition on degrading the subject to a mere
    network element).

7. In hazardous situations that prove to be unavoidable,
    despite all technological precautions being taken,
    the protection of human life enjoys top priority in a
    balancing of legally protected interests. Thus, within
    the constraints of what is technologically feasible,
    the systems must be programmed to accept damage
    to animals or property in a conflict if this means that
    personal injury can be prevented.

8. Genuine dilemmatic decisions, such as a decision
    between one human life and another, depend on the
    actual specific situation, incorporating "unpredictable"
    behaviour by parties affected. They can thus not be
    clearly standardized, nor can they be programmed such
    that they are ethically unquestionable. Technological
    systems must be designed to avoid accidents. However,
    they cannot be standardized to a complex or intuitive
    assessment of the impacts of an accident in such a
    way that they can replace or anticipate the decision
    of a responsible driver with the moral capacity to
    make correct judgements. It is true that a human
    driver would be acting unlawfully if he killed a person
    in an emergency to save the lives of one or more
    other persons, but he would not necessarily be acting
    culpably. Such legal judgments, made in retrospect
    and taking special circumstances into account, cannot
    readily be transformed into abstract/general ex ante
    appraisals and thus also not into corresponding
    programming activities. For this reason, perhaps more
    than any other, it would be desirable for an independent
    public sector agency (for instance a Federal Bureau for

the Investigation of Accidents Involving Automated
Transport Systems or a Federal Office for Safety in
Automated and Connected Transport) to systematically
process the lessons learned.

9. In the event of unavoidable accident situations, any
    distinction based on personal features (age, gender,
    physical or mental constitution) is strictly prohibited. It
    is also prohibited to offset victims against one another.
    General programming to reduce the number of personal
    injuries may be justifiable. Those parties involved in the
    generation of mobility risks must not sacrifice non-
    involved parties.

10. In the case of automated and connected driving
    systems, the accountability that was previously the sole
    preserve of the individual shifts from the motorist to
    the manufacturers and operators of the technological
    systems and to the bodies responsible for taking
    infrastructure, policy and legal decisions. Statutory
    liability regimes and their fleshing out in the everyday
    decisions taken by the courts must sufficiently reflect
    this transition.

11. Liability for damage caused by activated automated
    driving systems is governed by the same principles as
    in other product liability. From this, it follows that
    manufacturers or operators are obliged to continuously
    optimize their systems and also to observe systems they
    have already delivered and to improve them where this is
    technologically possible and reasonable.

12. The public is entitled to be informed about new technologies and their deployment in a sufficiently differentiated manner. For the practical implementation of the principles developed here, guidance for the deployment and programming of automated vehicles should be derived in a form that is as transparent as possible, communicated in public and reviewed by a professionally suitable independent body.

13. It is not possible to state today whether, in the future, it will be possible and expedient to have the complete connectivity and central control of all motor vehicles within the context of a digital transport infrastructure, similar to that in the rail and air transport sectors. The complete connectivity and central control of all motor vehicles within the context of a digital transport infrastructure is ethically questionable if, and to the extent that, it is unable to safely rule out the total surveillance of road users and manipulation of vehicle control.

14. Automated driving is justifiable only to the extent to which conceivable attacks, in particular manipulation of the IT system or innate system weaknesses, do not result in such harm as to lastingly shatter people's confidence in road transport.

15. Permitted business models that avail themselves of the data that are generated by automated and connected driving and that are significant or insignificant to vehicle control come up against their limitations in the autonomy and data sovereignty of road users. It is the

vehicle keepers and vehicle users who decide whether their vehicle data that are generated are to be forwarded and used. The voluntary nature of such data disclosure presupposes the existence of serious alternatives and practicability. Action should be taken at an early stage to counter a normative force of the factual, such as that prevailing in the case of data access by the operators of search engines or social networks.

16. It must be possible to clearly distinguish whether a driverless system is being used or whether a driver retains accountability with the option of overruling the system. In the case of non-driverless systems, the human-machine interface must be designed such that at any time it is clearly regulated and apparent on which side the individual responsibilities lie, especially the responsibility for control. The distribution of responsibilities (and thus of accountability), for instance with regard to the time and access arrangements, should be documented and stored. This applies especially to the human-to-technology handover procedures. International standardization of the handover procedures and their documentation (logging) is to be sought in order to ensure the compatibility of the logging or documentation obligations as automotive and digital technologies increasingly cross national borders.

17. The software and technology in highly automated

vehicles must be designed such that the need for an abrupt handover of control to the driver ("emergency") is virtually obviated. To enable efficient, reliable and secure human-machine communication and prevent overload, the systems must adapt more to human communicative behaviour rather than requiring humans to enhance their adaptive capabilities.

18. Learning systems that are self-learning in vehicle operation and their connection to central scenario databases may be ethically allowed if, and to the extent that, they generate safety gains. Self-learning systems must not be deployed unless they meet the safety requirements regarding functions relevant to vehicle control and do not undermine the rules established here. It would appear advisable to hand over relevant scenarios to a central scenario catalogue at a neutral body in order to develop appropriate universal standards, including any acceptance tests.

19. In emergency situations, the vehicle must autonomously, i.e. without human assistance, enter into a "safe condition." Harmonization, especially of the definition of a safe condition or of the handover routines, is desirable.

20. The proper use of automated systems should form part of people's general digital education. The proper handling of automated driving systems should be taught in an appropriate manner during driving tuition and tested.[57]

Germany's set of rules isn't perfect, but it's a damn good start. I know that's a lot to read, and I wouldn't be shocked if most of you just skipped or skimmed it, so here are the important takeaways:

- It's suggested that robotic vehicles are actually an ethical imperative if they prove to be safer than human-driven cars, although they also admit that forcing people to use them is ethically questionable (rule 6).

- Human life has precedence over animal life or property damage (rule 7).

- All human lives have equal value, and no specific information about the people can affect their comparative "worth." This is good to hear, as it addresses my recurring nightmare that robotic vehicles will decide if you're worth saving based on your credit score (rule 9).

- Situations like the trolley problem are addressed in several of the rules (rules 5, 8, and 9) and it's suggested that when these dilemmas do arise, an independent agency should be created to evaluate the situations and attempt to "process the lessons learned."

- Semiautonomous systems that require a human to instantly take control of the vehicle in emergency situations (Level 2 autonomy) are clearly discouraged, and possibly banned (rule 17). As you can see, Germany agrees with my assessment in Chapter 4 that semiautonomy sucks.

- Liability and accountability must be clearly communicated to a vehicle's passengers (rule 16) and in the case where the car is under autonomous control, liability is governed by the same sort of rules as in any product liability situation (rule 11); that means the company that made/sold the car is responsible.

- Regarding the possibility of malicious hacking, the manufacturers of the robotic vehicles are responsible for keeping their vehicles secure and safe as much as possible and as actively as possible (rule 14). The rule actually suggests that automated driving is only justifiable if the vehicles can be made secure, IT-wise. Ideally, this means companies face severe penalties or even outright banning if they cannot keep their systems secure.

It's complicated, right? Nothing's easy. While these rules are certainly a good start and are by far the most comprehensive set of guidelines from any government agency so far, there's still a lot of holes, and the problem is that those holes exist because the world itself just isn't that great at following rules. Even if we can manage to get every single robotic vehicle on the road to play by the same set of rules, the streets are still crammed full of things that don't give a brace of bowel movements about those rules, or, often, rules in general. There are times when, in order to operate within the messy, confusing muck of the real world, even a robot needs to know when to break or at least bend some rules.

Think about this: have you ever been to a big, dense, pedestrian-heavy city like New York, Hong Kong, or Mumbai? Sure you have, or you're probably sitting near someone who has, or, hell, just go on YouTube and watch some videos if you need to. Think about what crosswalks are like in those types of cities, what a typical Manhattan crosswalk is like in the middle of the day. The entire area is crammed full of people walking, biking, scootering, Segway-ing, sashaying, lumbering, loping, jogging, everything. There are also cars, taxis, and trucks everywhere, filling the street. While there are traffic lights and walk/don't walk signals, the use of the walk/don't walk signals is usually optional, at best. If people want to cross the street, they do. They look out for cars approaching them, and if they decide they have even a remotely plausible chance of making it across the street alive, they take it, relying on the fact that most drivers aren't interested in the hassle of running them over.

This means that a car stopped in front of a busy crosswalk should not rely on the signal lights to permit it to cross. If the car isn't in motion, people will cross right in front of it. In order to get through the crosswalk, a driver has to put the car in motion even while people are all around, and some may be about to cross, or in the process of crossing, right in front of the car.

The usual method involves inching the car slowly forward, not fast enough to actually cause a real threat to anyone, but with enough will and determination to convey the idea that they should get the hell out of the way, otherwise that driver's patience will soon be sucked dry and the car will propel itself forward with enough force to ruin your day. In short, to get through, you have to be assertive, determined, and a bit of an asshole.

The idea that a carefully programmed robotic car will be willing to potentially place a pedestrian in danger by slowly driving toward them seems insane, but if that car is to have any hope of being motile, that has to happen. It's not technically following the written rules of the road, but it is an adaptation to the unwritten rules of how things go in the big city.

Making things even more difficult is the fact that these unwritten rules are extremely regional, and every major metropolis seems to have its own dialect of driving and its own set of unwritten rules. In Los Angeles, for example, there is an extremely consistent and rigid unwritten rule about turning left at an unprotected traffic light, that is a left turn at an intersection with no provision for a green arrow traffic signal.

The Los Angeles rule is that when the light goes from green to yellow to red, up to three cars waiting to turn may turn on the red light. I lived in Los Angeles for more than seventeen years, and this rule was one of the most consistent things in my life there. Every Los Angeleno seemed to know about the three-cars-on-a-red rule, and when I described it to anyone else in the country, they looked at me like I was an idiot. And not the usual kind of idiot I'm assumed to be, a *dangerous* idiot.

Should a robotic vehicle follow this unwritten LA traffic rule? It's technically illegal, but in practice it's the norm, and not acknowledging the rule could potentially create more issues than just going with it would. I know if I was in the middle of an intersection when the light went red and some stupid robo-car in front of me refused to make the turn, it'd drive me batshit. I don't think I'm the only one.

That's pretty minor considering the driving habits and cultures of many other cities. I've driven in India, in Delhi, Mumbai, and Bangalore, and the driving there is closer to the movement of blood cells flowing through veins than it is any other street traffic I've seen. It's chaotic and terrifying and sometimes unpredictable, and yet somehow it flows and mostly works.

Lane markings are routinely ignored; cars drive into oncoming traffic if the road is smoother there; people and animals are always darting into traffic lanes; motorbikes and auto-rickshaws are constantly squeezing around and between lanes of traffic; and if there are traffic rules that are supposed to order this madness, they're about as respected as a Chuck E. Cheese's would be in Asgard.

How is a robotic vehicle expected to behave in such a situation? If it follows traffic rules to the letter, it'll be forced to remain as stationary as a statue. Should it work within the vernacular traffic rules, as it can best interpret them? Should it simply operate on a streamlined algorithm that has it plot a course to get to its destination as quickly as possible, without impacting anyone? Should there be special lanes and routes established for the handicap of rule following? I really have no idea, and I'm not sure anyone else does right now, either.

There are other reasons why we might choose to break rules or perhaps be a bit unsafe. It's very easy to make safety the top priority over every decision a robotic car makes, but the truth is we don't always want that in our cars. Sometimes we make a semi-informed decision to sacrifice some safety in the name of something else we want: such as convenience, excitement, fun, novelty, or emergency situations. This is just part of what makes being a human so confusingly wonderful, and just because it's not entirely rational or reasonable doesn't mean that it must be dismissed out of hand.

Sometimes this deliberate disregarding of safety can be trivial, like when a bachelorette party, all drunk and crammed into a limousine tooling slowly down the Vegas strip, sends the bride-to-be and some of her drunken pals up through the sunroof to sit on the roof, to yell filthy things at passersby and spill champagne all over the vinyl top.

Sure, it's a cliché, it's silly, and it involves a certain (very mild) element of danger, but it's also something that people really love to do, and will likely still want to do well into our possibly automated future. Would a robotic limousine allow such behavior, even though it quite clearly violates a number of traffic laws?

What about speeding, if the road and conditions are determined to be safe? Or loading an object so large into the car that it extends out a window or tailgate? What about an illegal U-turn on an absolutely empty street in the middle of the night that will save you a solid thirty minutes of driving? Can computer brains be programmed to be able to determine when violating a traffic law is harmless and offers a net benefit to the car's occupants?

As of now there are no real answers to this question, and, as far as I know, no developer of autonomous vehicles is putting much effort into designing lawbreaking systems for their cars, for some pretty obvious PR reasons.

Let's think about the opposite situation for a moment, though. We're trying to develop robotic cars that drive better than we do, that obey traffic laws better, and that are safer and more efficient. We want them to be all the things we struggle to aspire to, driving-wise. So, based on that line of thought, should they be more heroic[58] as well?

By "heroic," I mean it should be possible to effectively force robotic vehicles to undertake actions that, while placing the vehicle itself in harm's way, could help save people's lives.

I think there will definitely be some degree of altruism inherent in robotic vehicle behavior, such as automatically getting out of the way for emergency vehicles like fire trucks and ambulances, but I'm thinking of something even more involved. Here's what I'm imagining:

Let's say we're at a point where robotic vehicles are developed enough to be on the market, and that 20 to 25 percent of cars on the road are autonomous. These robotic vehicles are constantly connected to one another and the internet, as expected, and when they're parked, idle, or otherwise empty and unused, they can let local law enforcement and emergency services know, along with important details like their location and effective fuel/battery range.

Okay. Now, let's imagine something really terrible happens, like some deranged terrorist rents a box truck and is heading to a crowded part of town with the intention of plowing into a bunch of people, much like we've unfortunately seen many times before, like in April 2018 when a terrorist named Alek Minassian for the idiotic reason of not being able to get laid drove a rented van into a crowd, killing ten people.

These types of vehicular attacks can be brutal and alarmingly effective. But what if the attack could be stopped or curtailed by the use of robotic vehicles in the area? In this situation, it could be possible for law enforcement agencies to effectively commandeer nearby robotic vehicles and send them to help. They could help by parking between people and the threat, acting as barriers against vehicular attacks or even providing cover in cases of attackers armed with firearms or other weapons. Robotic vehicles could be used to attempt to ram a weaponized vehicle or force it off the road, and to block it to prevent the driver from using it to cause more harm.

There are many less action-movie-esque ways that commandeered robotic vehicles could help; they could act as widely dispersed eyes to track suspects' cars, or even follow cars of potential kidnappers or other potential criminals. They could form quick roadblocks to keep people out of dangerous areas, or they could be employed to help evacuate people from areas that are on fire, flooded, structurally unsound, or have any number of other problems.

Really, robotic vehicles could act as an emergency logistics fleet for a community, and could potentially do a lot of good.

Sure, there are plenty of issues here—it's essentially police commandeering private property, and that property has a high probability of being damaged or destroyed. It's a lot of power to give law enforcement at a time when trust in that institution has eroded quite a bit among many people and for good reason. Plus, having such a command channel at all leaves open the possibility that it could be hacked, and potentially by people you very much don't want commanding an army of robot cars.

There'd have to be some sort of reasonable compensation system set up for the owners of the cars that become ad hoc RoboCops, and I'm sure that system will be full of issues, but the general idea of police commandeering private property has been around for a while. The laws are known as posse comitatus statutes, and the idea that government officials need to pay for equipment or vehicles that have been damaged in official, commandeered use is something that's been considered since at least the Civil War. This is analyzed on the thoughtfully titled website the Straight Dope.

> [W]hat if they destroy or damage my property? That's less
> clear. In *United States v. Russell* the Supreme Court was
> faced with a claim for three steamers commandeered by
> military authorities during the Civil War. The Russell court
> found it obvious that "the taking of such property under
> such circumstances creates an obligation on the part of the
> government to reimburse the owner to the full value of the
> service." The court continued, "private rights, under such
> extreme and imperious circumstances, must give way for
> the time to the public good, but the government must make
> full restitution for the sacrifice."[59]

It's entirely reasonable for someone who owns a private robotic vehicle to not be crazy about the idea that the police or fire department may decide to take control of it in the middle of the night sometime. That's why this may be best handled in the same way we handle organ donation now; there's a system set up for it, it can save lives, no question, but people can opt in or out as they see fit.

For fleets of robotic vehicles owned by businesses, I think local municipalities could have the option of making such opt-in vehicular heroism required if the company wants to do business in a given community. Chances are that fleet vehicles will be the most common and available robotic vehicles, at least at first, and it feels easier to accept the potential destruction of, say, a robotic UPS truck than someone's personal car.

I know it seems sort of odd to demand self-sacrifice and heroism out of a machine, but if we're going to have machines replacing our jobs, why not throw hero into the mix? It never hurts to have some of those around.

CHAPTER 7

THEY SHOULDN'T LOOK LIKE CARS

THE WAY WE CHOOSE OUR CARS IS, IN MANY WAYS, CLOSER TO the way we choose our clothes than it is anything else. A car is generally the largest and most expensive self-contained piece of machinery we're ever likely to own, but the way we decide which one to get is far more irrational than the way we pick out some other large complex machine we'd buy, like a dishwasher or a riding lawnmower. That's because the way the car looks—and what that look conveys about us—is incredibly important to us.

Even people who claim not to care at all about cars, and make this claim with the same showy confidence that people who don't technically own televisions like to shout, I can guarantee you that you could find a car that they would absolutely not want to be caught dead in.

Your friend who drives a battered 2002 Toyota Corolla and takes pride in how infrequently they wash it most likely would not feel comfortable in a metallic purple Hummer with gigantic twenty-four-inch chrome spinner wheels. This means that they absolutely *do* care about what they drive, and that they've associated part of their identity with their car, as almost every car owner does.

There's so much of ourselves and our self-image, both actual and desired, tied up with our cars that the idea of a radical reshaping of the car to better accommodate the realities of robotic vehicles is an idea that is going to prove to be much more difficult than a completely cold, rational view of the problem would suggest. Cars are fashion, culture, status, and so many other things beyond the most efficient packaging of details like passenger area, cargo area, and drivetrain, and so much of the design vocabulary around cars is based on how we understand cars today. Autonomous cars very likely will diverge dramatically from human-driven cars, as they absolutely should, since the way we'll be interacting with them is so fundamentally different.

There are some significant practical design criteria that will be associated with autonomous cars, but before we get to that, it's worth talking about the less rational parts of automotive design.

Take performance cars, for example. There are some very classic sports car design motifs, including the long hood/short rear deck proportions, large wheels, and a certain powerful muscularity. Cars designed for performance have a look that reflects this, some of it practical, such as aerodynamic aids—like spoilers, wings, and skirts—as well as large brake rotors, wide tires, substantial air intakes, and venting. You know this stuff when you see it, and often these elements are exaggerated for effect. The overall look is serious and capable, and the car's aggressive design is meant to convey that not only is the car fast and nimble, but the driver *must* be skilled and capable of handling such a beast. A performance car suggests status via wealth and status by appearing to be someone who has the ability to handle such a raw, untamed beast, whether that is true or not.

So what happens when the driver is out of the equation? When a car is capable of driving itself, the skill of the driver ceases to be a factor, so what, then, is the point of an autonomous performance car? Could such a thing exist? Maybe. We'll come back to the concept of autonomous car "personality" and behavior soon, but first we should discuss the physical design of robotic vehicles.

Conceptually, the dominant thematic trend in auto design for the past two decades or so has been aggression. The overall tone, the look and feel, the general visual character of car design has been pretty steadily growing more and more angry, intimidating, and imposing. While this was once a phenomenon primarily limited to sports and muscle cars, the trend of making cars look like angry alien vacuum cleaners has disseminated throughout every automotive segment.

Even small entry-level cars that, in decades past, may have carried a friendly, eager look, now all have faces that suggest they may want to try and shiv you. Consider a car like the Chevrolet Spark, one of GM's most affordable cars. The Spark was derived from a car, the Daewoo Matiz, built by GM's Korean division. Years ago, the Matiz looked like the car on the left. The current-generation Spark is on the right:

The new Spark is pretty mild, aggression-wise, when compared with many other cars, but it has all of the key modern aggressive design traits, the most notable of which are angry-looking headlamp-eyes, and a large, gaping grille (even if most of it is fake). The older Matiz is at the exact opposite end of the spectrum, with a "face" deliberately designed to look anthropomorphic and friendly.

There are some exceptions to this, of course, but the exceptions are almost always modern reinterpretations of vintage designs, like the new Beetle, the reborn MINI Cooper line, and modern Fiat 500s. Their original designs were from an era when it was acceptable for a car to appear to be a willing and cheery companion; now, as studies have shown, people tend to prefer angrier, more aggressive looking cars.[60]

Why is this, exactly? It seems to come from some desire of drivers to appear powerful and intimidating to other drivers, a reaction possibly stemming from insecurity and fear about the potentially dangerous action of driving, or a reaction to the hostilities and frustrations many drivers experience, especially in areas of high traffic congestion. It's also possible that the perception of being powerful provides a status boost similar to what people seek from other status-signaling automotive traits like perceived cost and exclusivity.

Now, all of this only really makes sense when we talk about human-driven cars, since the inherently prosthetic nature of cars makes most people interpret the car as a sort of extension of the driver. So when you see an intimidating, angry looking car, the desired result is that you, at least on some unconscious level, ascribe those same traits to the driver. But what happens when there isn't a driver? Are people going to want angry and powerful looking cars that drive themselves? The entire context of the look of the car changes once we remove the driver from the equation. A person driving an aggressive looking car may appear intimidating, at least on some level (the owner hopes), but our knowledge that a human being is in control of the thing tempers our reaction. Centuries of human culture have given us at least some sense of understanding about the driver's possible motives, and we all have, collectively, enough trust in human society and the rule of law to accept that, most likely, this person means us no harm.

When we encounter a vehicle that goes too far beyond our accepted norms, things change. If, for example, you saw a car that was a dead ringer for something out of *Mad Max*, full of spikes, spears, and jagged rusty bits, you might be more nervous.

Well, now that I really think about it, so much else of our culture would probably keep you from worrying, and instead you'd just assume it was done for artistic purposes or is part of some movie or promotional thing. Really, it's pretty difficult to scare a modern driver through the design of a car.

But once we take the driver out of the equation, I think there's more harm that can be done. A robotic vehicle designed to look aggressive is likely a very bad idea, at least from a public acceptance standpoint. We don't have centuries of experience to condition us to trust robots. In fact, most of the popular culture surrounding robots contains a significant number of stories designed to elicit fear and distrust. A large robotic vehicle with a face that looks like it wants to try human bone marrow as fuel is absolutely going to have an effect on people who interact with it, regardless of how much they may know, rationally, about the safety protocols and myriad reasons why the machine has no interest in harming them, and how counterproductive for everyone involved a murderous rampage would be.

It's because of this that I wouldn't be surprised to find out that at least the first few generations of robotic vehicles will also likely mean the end of hyperaggressive automobile design. We may even see a bit of a pendulum swing in the opposite direction, with robotic vehicles designed to not just be unintimidating, but actively cute. In fact, we're already seeing it; Google's first fully autonomous experimental vehicle resembled, famously, a koala bear, one of the most notoriously cute animals known in the natural world.

Google deliberately designed these cars to be very cute and approachable. YooJung Ahn, the head of design for Google's self-driving car program, couches the design in terms of safety:

The exterior shape also speaks to this concept of safety. Its soft, round edges are coated in foam and the windshield is made from polycarbonate. It bares [sic] a distant resemblance to the circuitous shape of the first VW Beetle or a '50s-era Fiat 500, cars often described as cute. The sensors and lights on the Google prototype don't feel sterile, stark, or sharp; there's depth and fluidity across the surfaces. That's why its prototype is like the little white bubble car that could.[61]

Cute is disarming. Cute puts people at ease, evokes a sense of youth and harmlessness, and triggers something in people to make them want to help, not hinder, whatever the cute thing is. Robotic vehicles, especially the first generations, will desperately need these sorts of reactions from people. They will require trust and patience and, occasionally, assistance. They will need help, and as a way of helping to ensure they get it, they're probably going to be at least a little bit cute, or at least neutral enough to not be intimidating.

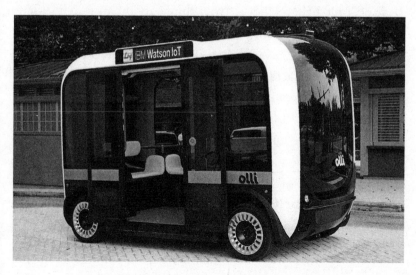

Another example of a robotic vehicle designed to be cute, Olli, comes from a start-up company that is making autonomous buses for set routes. The buses have a distinctly neotenized look about them: huge forehead, friendly face, rounded edges, and so on. They sort of look like wheeled baby orca whales. There's nothing scary about them, which is entirely the point. A cute robotic vehicle is as capable of failing and crushing you to chili as is a vehicle that looks like Cthulu with a hangover, but you're a lot less likely to expect it, and that's a crucial distinction that may very well be the key to early acceptance of robotic vehicles.

In thinking about this, it's important to keep in mind a few unique properties of cars: that they're both a means to get from one place to another and, simultaneously, a "place" of their own, as we've discussed, and also that a car's status as either public or private space is by no means clear. A car out in the middle of a city may seem like a public space, being a heavily windowed box out in the open for everyone to see, but most of us treat the interiors of cars like private spaces no matter where we are. If you don't believe this is true, think about how many times you've seen (or performed) acts of vigorous nostril-spelunking or sung to yourself like an idiot, or done any number of potentially disgusting or embarrassing things in your car with the calm confidence that only comes when you think you're alone.

In reality, if you're nose picking or singing along to Smash Mouth or doing something equally embarrassing while stuck in traffic on I-405, you are about as un-alone as you're ever likely to be. At the same time, I freely admit I'd treat my car as a private space in that context, too, and I suspect most people would. The interior of a car is a more important location in our lives than we tend to realize, and will only become more important with the coming of robotic vehicles.

The most important thing to realize is that the conventional, mainstream automotive design of a box for an engine, followed by a box for passengers with two rows of forward-facing seats, followed by a box of some proportion for cargo at the rear doesn't really make sense when it comes to autonomous vehicles. The idea that everyone should be facing forward, with attention focused on the large window at the front of the car only really makes sense if a human is driving the car. It's not a layout you'd pick for almost anything else, except perhaps a very small movie theater. When people are together in other contexts, they like to either face one another, or have surfaces to work on, eat on, play on, or whatever. In short, people like to spend time in rooms. In fact, I bet many of you reading this are in a room at this very moment. I'd also bet that you've done all manner of things in rooms, and many of those things are activities you'd choose to perform while being whisked around in a robotic vehicle. A robotic vehicle is something that does not require a human to pilot it, and, as such, has no need to be designed to be a prosthetic motion device for humans.

Robotic vehicles should be designed from the inside out, and those insides should contain the maximum volume of space possible for the vehicle's size. The best way to do that is with boxy shapes, which means that the ideal baseline form for a robotic vehicle would be a van.

Yes, a van. A van is, essentially, a room on wheels, and in a vehicle that's intended to be ridden in and never driven, that's what people will want. The fundamental idea that the interior of a car should be a flexible, reconfigurable space suitable for a variety of activities beyond sitting and facing forward, like you would while enduring a boring religious service, has been around for quite a while. Perhaps the first really serious application of this concept for an everyday car—not a camper or anything like

one—came around 1932, the work of an innovative automotive designer named William B. Stout. Stout is most famous for designing a very advanced, elegant Zeppelin-like vehicle, the Scarab.

The Scarab was designed as one long Art Deco-looking streamlined box, with a large V8 engine mounted at the rear. The passenger compartment could be configured fairly conventionally, and sit a good number of people, or, if you were bold, be arranged something like a little living room, complete with facing seats and a flip-out table.

Of course, the Scarab was decades away from anything even approaching autonomy, but, suitably converted with the necessary electronics and actuators, would be a fantastic design for a robotic vehicle.

The Scarab wasn't longer or wider than conventional cars of the era, it just used that same volume of space with much more efficiency, perhaps a result of Stout's background in aircraft design. It hinted at the idea of a minivan decades before the first real ones made it to market. It was a brilliant triumph of packaging, so, of course, it failed to catch on.

The Japanese have always embraced the idea that a car should be an enclosed volume of mobile, personal space, and as such they have a long history of developing all sorts of fascinating van-like car designs. Some of these incredibly space-efficient designs come from Japan's kei-class regulations: a set of parameters that define small city cars for Japan's highly congested cities. These cars have strict size requirements and engine requirements (maximum 660 cc and 64 horsepower, if you're curious), and many of these cars are designed to use as much of their allotted space as possible, which results in some novel-looking "tall boy" micro-van designs, genuinely tiny cars that have a striking amount of usable space inside.

The reasons why cars like these thrive in Japan—and pretty much only Japan—are interesting, especially in the context of how people live in very dense metropolises. Personal space in a city like Tokyo is at a premium, especially private personal space. A car becomes something more than just transport in this context; it becomes a mobile bubble of private space that may be incredibly hard to find otherwise.

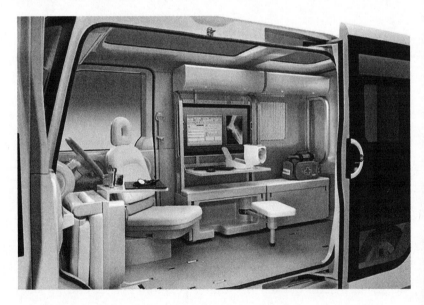

Japanese carmakers have long understood the role of a car as a personal space, and have, at some points, even built and marketed cars with privacy as a primary focus. One of the most infamous examples of this was the Honda S-MX, which was designed to be, at least in part, a "mobile love hotel."[62] Yes, that's right. Honda once built a car specifically for people to bone in. They weren't even especially coy about it. The seats folded into a roomy and comfortable-looking bed, and their ad campaign was very, um, love-focused. Here's a bulleted list of features from the brochure, as clumsily translated by machines:

- friendship that cause if you fall, love that fall together. (Full flat sheet)

- one step of love would casually begun. (One-step floor)

- swell love protect Futari. (SRS airbag system)

- of love that two of the heart become one. (2.0ℓDOHC engine)

- Love pupil is always shining. (Large headlights) is called the love of the things which are not seen only (Panorama side window)

- love wants to look at the shoulder shifting same future. (Twin bench seat)

- Although the entrance is one of love, and there are two of marriage and heartbreak the outlet. (ABS)

- Pull in love, and then press in love (real-time 4WD)

- that shake the love like music chest (delicatessen Sea suspension)

- love color is the color of the flame of fire. (Lowdown specification)

- love begins familiar place. (Honda Clio store)

I mean, that's pretty clear.

Japanese concept cars have also long explored the idea of the car as a mobile, multipurpose room, and there are many Japanese concept cars that I think we will one day consider to be the conceptual ancestors of robotic vehicles.

Sometimes they focus on flexibility and practicality, often with very specific interior setups and designs. Other times they play with ways that a box can be worked to feel sporty or aggressive or sleek or fun, even despite the traditionally non-stylish proportions. I've always thought Honda's 1999 concept car, the Fuya-Jo, was a good example of this. It was described by the automotive press as a "party on wheels"[63] and while, sure, a lot of the design detailing was cloying and overdone, the basic concept is strong, a dramatic take on a box on wheels that does not feel boring or utilitarian.

These concepts bring us back to our earlier question: what's going to be important or compelling to people when it comes to the styling of cars they will never drive? I'm mostly thinking of private robotic vehicles, though design and a distinctive style will be important to fleet vehicles as well.

A safe bet is something that has always been important to car buyers: status. A car that telegraphs how much disposable income you have, accurate or not, has always been in demand. Many people like cars that suggest wealth, status, or power. This rather vain and, objectively, silly reason is why huge SUVs have become the default car type in America, and much of the rest of the world today. As soon as celebrities and VIPs of various types—from the president's ichthyologist to pop stars—started being ferried around in convoys of black SUVs, sales of luxury SUVs, usually in black, rose accordingly.

In case you're somehow still not convinced of how potent a force it is in automotive design, despite everything we've already discussed about status, consider the case of arguably the least status-conscious car: the Tata Nano. The Nano is the cheapest new car in the world,[64] coming in at about 2,700 American dollars. Really, it's an engineering marvel: a very usable, four-door car for that cheap is an engineering triumph. The goal of the Nano was to provide a way for poor Indians who drove around on mopeds with their entire families a much safer, weatherproof alternative. Technically, the Nano absolutely succeeded at this.

The Nano never sold nearly as well as expected, though, because even though the car delivered exactly what was promised, it was marketed as the cheapest car in the world, and nobody, not even the poorest people in the world, want to be seen in the cheapest car in the world. Objectively, this is madness; there's no way a Nano isn't a dramatic improvement over four people crammed onto a moped in the rain. But once branded with the stigma of being the cheapest, nobody wanted to be seen in one. It's not rational, it makes no sense whatsoever, but it's very real and very powerful. That's why I'm pretty sure the first private robotic vehicles sold will be expensive, and look like it. Initially, I think the perception of status will even outweigh the need to make the vehicles friendly and approachable-looking. If we accept that driverless vehicles make the most sense when designed to maximize interior space, forcing a boxy, van-like general form, then the detailing, materials used, and known brand/status associations will become even more important.

This actually works well for an interior-focused vehicle, as there's already a lot of experience building luxury-focused vehicles for wealthy people who never intend to drive themselves. From the earliest chauffeur-driven town cars to modern Mercedes Sprinter van conversions that are designed to replicate private jet interiors, for decades the rich have been exploring how to enjoy cars without actually driving them.

More fleet-oriented vehicles, such as robotic vehicles operated by companies like Uber or Waymo, or whomever else may eventually get into this space, will likely have more of a public-transportation-like feel to them since, really, they are private public transportation vehicles.

Of course, if all of this does eventually work, there will be basic autonomous public transportation, and there's not much reason to think they won't generally be buses as we already understand them. Buses are already big boxes full of people, and for the goals of a mass transit vehicle, that's pretty hard to beat.

Now, things get more interesting when two things happen: first, we start to consider robotic vehicles more as robots and second, we realize that this means we don't always have to be tagging along. I'm talking about an entirely new class of automotive design here, cars that are designed to do specific jobs and never have human passengers.

Up until this point, large-scale robots comparable to what I'm imagining have been limited to, primarily, industrial uses, and people very rarely encountered them in daily life. This would be different. These robotic vehicles could be incredibly common, sharing the road with robotic passenger vehicles and human-driven vehicles, and could be owned by businesses, governmental or municipal agencies, or even private owners. In fact, I think it's possible that for privately owned robotic vehicles, it may be more common for a family to have a robotic vehicle not designed for passengers long before they have one they themselves ride in.

I mentioned the concept of a cargo-only vehicle a few chapters ago, but it's worth looking at its design in more detail. A cargo-only autonomous vehicle, essentially an errand-running robot, would essentially be a wheeled cargo box. As such, it could be much smaller than a conventional car, about the size of the bed of a pickup truck. In fact, the cargo area should be about this size, to allow for standard sheets of plywood and other commonly purchased and standardized larger-scale objects. The design of such an errand-bot could be pretty basic, really:

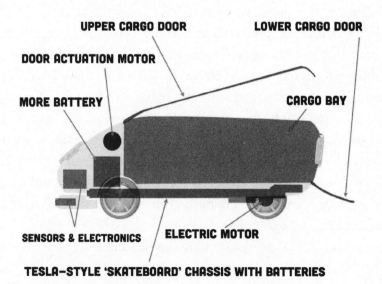

UPPER CARGO DOOR   LOWER CARGO DOOR

DOOR ACTUATION MOTOR

MORE BATTERY   CARGO BAY

SENSORS & ELECTRONICS   ELECTRIC MOTOR

TESLA-STYLE 'SKATEBOARD' CHASSIS WITH BATTERIES

In order for such vehicles to be really effective, there would need to be a commercial infrastructure designed to accommodate them, but, luckily, the requirements for that are pretty minimal. For example, consider what a grocery store would need to do to accommodate such un-peopled vehicles. You'd need a way to order groceries online, which already exists. Once ordered, the grocery store would need some staging area for the errand-bots to park. The bots would need to be able to communicate with the store when they arrive, and provide some sort of confirmation as to what order they are there to pick up. That could be as simple as instructing the car to text (or send via some other, perhaps automotive-standardized wireless protocol) a particular confirmation number upon arrival.

Once there, a normal, boring, moist human being could carry the groceries to the car, which would unlock and open its cargo door once the handshaking with the grocery store's system had taken place. After the cargo area is loaded, an integrated camera would take a picture of the loaded cargo bay and send that to the owner's phone. Once the owner confirmed the robot was loaded and ready (I suppose they could have a default to accept, if the owner is trusting and lazy), the errand-bot would be given the command to head back home, where it would arrive like any conventional car, and be unloaded just like people have been unloading groceries from station wagons for nearly a century.

Sure, more of this process could be automated, but that would require designing and building such things as docking systems, cargo conveyers, and automatic loading systems, when nearly all of the loading and unloading could be handled by human beings, who don't require any design or research and development to do the job.

Because all that would really be needed is a place to park

and a bit of communication back-and-forth, nearly any business, even small independent retail outlets, could accommodate such machines.

There are, of course, security issues involved. Errand-bots would be prime targets for thieves, as they're basically unattended rolling treasure boxes, potentially filled with expensive tablet computers or desirable pharmaceuticals or other valuables. Of course, they could also be filled with string cheese and tampons, so there would be no guarantee that hijacking one would even remotely be worth the risk.

As potential prey, an errand-bot would need to maintain constant communication with its owner. It would need to use live video streams to record all around the vehicle, and have the ability to call for help electronically if certain criteria are met, like attempted forced entry, the vehicle being lifted off the ground, and certain visual criteria from the cameras (such as being surrounded on all sides); the owner, watching the live stream from the errand-bot, could choose to call for police if they noticed something worrisome on the video, from feeds both inside and out.

As I mentioned earlier, these vehicles could be tasked with following your human-driven car as well, expanding your cargo capacity and acting as your own personal support vehicle. They could even tow a human-driven car that has experienced mechanical failure. As you may have guessed, many of these ideas have come from my own fevered fantasies as I've sat by the side of the road, desperately trying to coax some old beater back to life, wishing for a magical robot to bring me parts, beverages, and, when I eventually and inevitably give up, tow my sorry ass back home.

I've just been imagining privately owned autonomous cargo vehicles, but the truth is commercial applications of such vehicles will probably be far more likely. Small-scale robotic vehicles could prove to be an ideal last mile solution for companies like Amazon, which has experimented with drones and other similar exotic ideas. A robotic cargo vehicle could carry far larger packages than an aerial drone, and would be far cheaper to operate. Large trucks with long routes could be replaced with fleets of small, very easy to park unmanned cargo-bots, driving to people's addresses or a chosen location and then sending a text when they're ready for the customer to send their confirmation code and take their package.

Again, security would be an important issue, but these vehicles would be in constant communication with both customer and whatever company owns them. They'd be trackable and sturdily built to resist easy unauthorized access. I think they'd prove to be much more difficult targets than one would think.

Robotic vehicles not designed for passenger transport could be designed to accommodate a huge number of services that currently require specialized human-driven vehicles. Garbage trucks, for example. Most now have robotic arm systems to grab and dump trash cans; it's not a huge leap to imagine these slow, lumbering mastodons driving known, set routes automatically. A human attendant would still likely ride along, to help in case of minor equipment issues, unexpected situations, and dealing with people, but most of the actual work could be accomplished autonomously.

CHAPTER 8

THE DEATH OF
THE JOURNEY

THERE'S NO QUESTION THAT THE EVENTUAL COMING OF
self-driving vehicles—whenever that actually turns out to be—
will be something of a revolution in transportation and will absolutely bring considerable benefits to many, many people. People
who are unable to drive for medical or physical reasons will gain
dramatically greater freedom of travel, traffic fatalities will likely
drop by a considerable amount, and a great many mundane,
tedious, or even dangerous tasks, jobs, and other ventures could
be automated, freeing people for other pursuits.

I mean, everyone's pretty excited about the possibility of it
all; I'm even writing a book about it, and I'm pretty convinced
we're not as close to having these things on the road as many
people think. While there's considerable mistrust and hesitation
about the idea, generally the sense is that this is a revolution that
is going to happen, and people are eager to reap the rewards.

Like any revolution, it's not going to be entirely positive.
Sure, there are real, tangible benefits, but those benefits come at a
cost, and that cost is not always something that's entirely rational
or measurable, but significant nevertheless. I don't bring up these
issues from some Luddite-inspired, anti-progress point of view,
but rather as someone who appreciates the unique cultural and
emotional aspects that a century of human automobile driving
has brought humanity.

As we all know, the car completely changed the human landscape: how and where we live, how we build our cities, the range

of places we can travel, and so much more. The car created its own cultures and subcultures, and just as we changed the world when we brought automobiles into it, automobiles changed us as well, and some of those changes are worth acknowledging and perhaps even preserving.

One of the biggest, most subtle and yet least-considered changes that the potential coming of robotic vehicles will bring is what I'm calling, with more than a little bit of high school-level drama, *the death of the journey*. The introduction of the automobile brought something to humanity that was never really possible before: the ability to undertake a very long journey almost completely independently, and to remain involved and immersed in every aspect of that journey.

Sure, railroads allowed for long journeys before private car ownership was common, but there's a big difference there. A journey by train is communal, and the person taking the journey can only plan his or her voyage within the confines and restrictions of the railroad and its schedule and routes. You're getting a ride, essentially, on a large machine that would make the trip whether you were part of it or not.

This of course goes for air travel as well, and even bus travel—pretty much any mass transit-type system is the same.

During the trip, you're able to look out the windows and consider the area you're traveling through—well, I suppose on an airplane this isn't always the case, as most of the trips tend to have views out the windows that are either cloudscapes of considerable beauty but absolutely free of geographic context, or just featureless expanses of white or black. Even though you could look out a window, that's not how we generally spend long journeys on trains or aircrafts.

Most of us focus our minds on other things during these trips, reading books, watching movies, listening to music, playing games, sleeping, working, all that sort of thing. We focus our minds inward or on things that have nothing to do with the trip we're currently going on because, fundamentally, we're not really taking a journey, we're cargo, being shipped from one point to another.

Trips like these, on airplanes, for example, are really more digital than analog, in the sense that they're about two binary points: the origination point and the destination point. Time is killed in between, but it's not really related to the journey itself; you get on at one place, do whatever for a few hours, and end up in a completely other place.

That's not a journey. That's clunky teleportation.

A road trip in a car that you drive is entirely different. When you drive your car, you're on a very obvious and unignorable journey. You're in charge of the operation of the vehicle; as a prosthetic, you're amplifying your body's own ability to move, and you're physically moving down the road. You are causing these actions to happen, and it requires the use of your body and mind.

Road trips can be long and tiring because you're actually doing the work of traveling; you're not just cargo being whisked from point A to the newly gentrified point B—you encounter every single inch of the area between those two points as you look through your windshield while you control your car's speed and direction. This is a very visceral, real, and engaged way to travel. You understand the distances you're covering and can watch the terrain around you change from one climate to another. You can see deserts gradually start to moisten and take bloom; you can see a city as it grows out of a rural area, and then again as it fades away into farmland as you leave.

A road trip you drive is analog, in that it's not an on/off, here/there situation, it's a long string of distance and experience that flows from one place to another place. And it's only possible because of the car.

Sure, people used to travel by animal power, but that was inherently slow and the distance you could reasonably cover was pretty limited. All of a sudden, a car capable of going a mile a minute changes everything. A mile a minute is fast enough to cover some real ground—a day's worth of driving can get you about five hundred miles, about a fifth of the way across the entire North American continent. That's a significant amount of land, particularly when traveled at a speed that allows a driver to really see and experience what they're driving through.

Plus, when you're driving, there are stops for gas and food and seeing giant balls of twine, or the car Hitler supposedly once defecated in, or a merkin made of Sasquatch fur or any number of other things one is likely to find off the side of a highway. There's no way to drive through an area and not engage with it in *some* way, and that's a valuable thing.

This is also precisely the sort of thing autonomous vehicles will take from us. While there are many people who no doubt will be thrilled by the idea of not having to remain alert and awake on a long road trip, robotic vehicles will effectively be transforming the process of a road trip into something much closer to a trip by airplane, train, or bus. It'll just be another way we get in something at one place, sleep or surf the net or work or whatever, and then end up in another place. It's another form of travel that reduces the world to a series of destinations instead of an entire continuum of places, people, and experiences.

It's not just the physicality of the world that we'll lose, there's the time itself. I know this sounds counterintuitive; aren't vehicles we don't have to drive going to *give* us far more time? Time we can use for all sorts of more productive purposes? Yes, it's true, your time won't be spent driving, and, with most robotic vehicles likely having a connection to the internet, you won't even be out of touch with the larger world during your trip. You'll have the freedom to spend your time in so many ways other than driving, but you'll be losing one very important freedom: the freedom to *not* be productive.

Anyone who thinks that a vast majority of working adults' time in robotic vehicles won't be spent working even more is a beautiful, deluded simpleton who should probably get some hugs before they wise up to how shitty the world really is, sometimes. As soon as a six-hour road trip doesn't require you to drive, you can be damn sure someone will require you to do something else, something productive. With the advent of email, smartphones, and generally being in constant communication with everyone, all the time, the pressure to always be available and working is considerable. There's been whole books and studies and all manner of media devoted to just that concept. And while this has been going on, only two human activities commonly undertaken while awake have managed to remain sacrosanct and free from pressure to be productive: voiding wastes and driving. And only one of these is really something you can get away with doing for hours at a time.

Long drives have been the greatest champion of the freedom to not be productive, far more than anything else ever was and likely ever will be. In fact, driving, because of the potential dangers involved, has legal restrictions on how much you can focus on other things at the same time; you can't sit and look at your smartphone, even though a great many people try, because eventually you'll end up in a wreck. Driving requires enough of your brain and body to exempt you from really having to do anything else.

If you're driving, you get a pass from so many things: answering phone calls and texts, returning emails, reading things, watching things, deciding things, pretty much everything. I'm driving. Leave me alone.

What makes driving so magical, is that it doesn't require *all* of your brain, and the parts it doesn't need are free to really enjoy things. Things like conversations with whomever you're riding with, listening to music, stories, or podcasts, and, perhaps most importantly, letting your mind get into a sort of meditative state where you can really think about things.

Lots of people drive when they really need to think about something, because the process of driving requires just enough conscious focus to allow the rest of the mind to freely wander. Part of your mind is occupied; you're adjusting the throttle, you're steering, you're remaining alert for other cars or a misguided armadillo or whatever, but other parts of your mind are free to rest and wander. This wandering is where new ideas are often born, where realizations come about things that your mind is normally too occupied to find.

I don't think we generally realize how big a deal it would be to lose one of our few remaining refuges from having a boatload of things expected of us at any given time, but I think we will feel the sting of it once robotic vehicles become commonplace. People who take mass transportation now already have had to live like this for a long while, but even then there's a difference, especially if private ownership and/or use of robotic vehicles becomes common. A crowded subway or bus limits what can be expected of you. If you're alone in a car, it's different.

Maybe the amount of time people spend in bathrooms will increase when we stop driving ourselves places. It's pretty much the only private respite we'll have left, at least until that miserable day when that gets automated, too.

Freedom from productivity is not all that autonomous cars will take from us, in their relentlessly logical, unhostile way. Robotic vehicles, by their very nature, pose a threat to one of the most important and underappreciated aspects of driving a car. It's something that is subtle and perhaps even easy to forget about as robotic vehicles are being developed, but I guarantee it's something that everyone who has ever driven understands and has enjoyed at one time or another.

What's lost will be revealed with the very first act we will undertake when we decide to take a trip in a robotic vehicle: giving a destination. Every fully automated vehicle will require a destination before it'll move an inch, because without a destination, what's it supposed to do? The nature of any trip in a robotic vehicle is a goal-oriented trip, with the end goal being the destination you have in mind. After all, that's the whole point of transportation, right?

Technically, yes, sure, that's right. The whole point of not just autonomous cars, but regular cars, or planes, trains, buses, steamships, or whatever it is you use to get you from one place to another. But that doesn't mean that's all we've ever used vehicles for. I'm just about certain that every driver reading this has, at some point, gotten in his or her car and driven with no idea of the destination. The reasons we do this are many and quite varied; sometimes we drive for the sheer love of the physical experience of it; sometimes we drive because the act itself has an effect on our minds, allowing us to think, calm down, or evaluate things, or any number of other, often quite personal reasons that are very specific to ourselves and our relationship to cars and driving.

Sometimes we drive without a clear goal out of a sense of adventure, or exploration. Sometimes we're seeking something, but we're not exactly sure what, but we know that we may be able to discover it by getting out there and actually *looking*. Sometimes we're hunting something specific but only half-remembered, making us mobile detectives, creeping through neighborhoods and trying to identify what is familiar.

We may drive for privacy with someone we wish to be physically close to, or able to talk with, in a private, intimate space, especially in situations and environments where privacy is a limited resource. We may drive to learn about a new city, or to tour, nostalgically, an old one. We may drive aimlessly to kill time or be out of contact for a set period of time. We may drive just because we love the way our car feels, sounds, and smells, and we like ourselves a little better when we're driving it. We may drive by places with reflective windows just to get a glimpse of ourselves in our car because for some stupid reason that's an image we never grow tired of seeing. We sometimes drive to wander, with no interest in a specific destination at all.

Artificial intelligence systems can do an incredible amount of amazing things, including interpreting and reacting to a chaotic world well enough to drive through it, but, so far, no artificial intelligence system is capable of originating any actual idea from nothing. AI systems are fantastic at taking algorithms and data and extrapolating to analyze or process in some complex way, but they're not going to, out of the blue, suggest grabbing some gyros for dinner unless they've been programmed with an algorithm to evaluate a number of factors to determine when may be the best time to suggest Greek food.

You're not going to be able to jump into a robotic car and tell it to just go for a drive and see what happens. Well, at least not without cheating by, say, having a set of preplanned routes that are randomly selected whenever anyone asks to "just drive."

The more I think about it, the more I'm starting to think that there will be some novel substitutes to give robotic vehicles the option of just "driving around." At the most basic level, you can imagine it wouldn't be hard to specify how long you want to be gone before ending up back at your original destination; the car could use random number generators to dynamically create a randomized path within some set radius.

I can imagine how the desire to aimlessly drive could be monetized and commoditized. Let's say your robotic vehicle offered a feature where it could just drive around, but that allegedly aimless drive was really a disguised mapping path to gather data for some company. Really, a savvy robotic car owner could have the option of renting the use of their car for mapping, surveillance, or any number of other location-based data-gathering purposes, and if you wanted to go along for the ride, well, why not?

Will this really be the same as just driving around on your own? I don't think so, but, then again, no one ever paid me to drive around aimlessly.

Losing the ability to just wander is a big deal. There may be ways to still permit this in a self-driving vehicle, though, and I do think this ability is worth having. A robotic vehicle that maintains conventional human driving controls, would, of course, be the simplest solution, but that assumes the human inside actually knows how to drive, something that can likely be counted on less and less as robotic vehicles become more common. A way to guide a robotic vehicle without actually having to drive it may be needed. There may be a way a robotic vehicle could be "guided" on the fly. Perhaps a route could be drawn with a finger on a map on a screen, and the car would follow the path to the best of its ability. Perhaps some kind of joystick, trackpad, or D-pad type of directional input could be used.

I know it seems a little strange to pilot a car using what is essentially a video game-type controller, but part of the deal with the widespread use of robotic vehicles is that we can no longer guarantee that the people inside a given car will have any idea how to drive a car, which means the only options for guiding a car in real time are ones where the car is still doing most of the work. In these situations, the human "driver" is really just sending the car a continually updating stream of micro-destinations, and the robotic car is using its normal driving/obstacle avoidance/safety systems to travel.

Interfaces like these would serve a number of uses beyond the intangible benefits of permitting wandering. Consider going to a large convention, concert, racetrack, or cult rally where once you get to the destination itself, a lot of driving is required to find parking, say in a large lot or an open field. It may be possible to automate looking around for parking, but how do you program a car's vision system to watch people walking in the lot and determine by their gait, what they're holding, their state of urgency, and many other subtle factors that only years of living as a human can help you notice, all of which help determine if they're heading to their car to leave or not? If you decide the person looks like they're leaving, great, follow them to their car and wait for a spot.

Can we even program a car to understand this sort of thing? I mean, half the time in this situation I usually mistake a fellow human's intent myself, leading to an awkward exchange as the human (it's feeling weird calling other people "humans" now, just so you know) gets something out of the car and then begins to return to the event on foot, and my eye contact and hand gestures that mean "hey, are you leaving?" are returned with half-apologetic shrugs, shaking of heads, and more pointing and waving gestures.

I'm pretty sure getting a computer to be able to understand all the subtle nuances of human behavior as viewed from the perspective of a vehicle is likely as big a project as getting a robot to be able to drive a car without crushing everyone around it to thick, chunky custard. Human behavior is complicated and usually deeply weird, and I think robotic vehicles are going to affect it in many more ways than we realize. In fact, I think when (and, I guess, if) we make a shift to mostly robot-driven cars, our fundamental perception of ourselves and our place in society will change.

Right now, the ability to drive is a hugely powerful component of our independence as individuals in a society. Large, dense cities with well-developed and well-used public transportation systems offer a vast amount of personal mobility freedom as well, but as an individual you're always reliant upon and at the mercy of organizations much larger than yourself. And sure, you can take the subway or bus almost anywhere in New York City, but if you want to leave the city, you're going to have to use a car—either by driving one yourself or getting someone else (taxi, Uber, easily manipulated friend, or unusually well-trained and capable dog) to take you there. No matter how good a city's public transportation system is, it's still not the same as having your own, independent car.

Now, I know many people consider private cars to be horrible environmental disasters that have created urban sprawl and soulless suburbs, and are awful, violent things that are destroying everything. While, sure, some of these things are no doubt true, that's a miserable and pretty limited way to look at cars and driving.

A personal car also offers one of the most powerful and fulfilling modes of mobility that human beings have ever enjoyed. Your own car means the ability to change your surroundings dramatically, to leave a bad situation and seek something new, to hunt down adventure or opportunity or whatever it is you want, wherever it is. Your own car means you can go where you want even if there are no subway lines, train tracks, or airport there. It's by far the most independent and unrestricted way to travel, and there's a feeling you get when you have a car and you know, if you had to, you could use it to escape, to break free, to find a new life.

For countless people in bad situations, the acquisition of their own car made a dramatic, incalculable difference. A car is freedom made physical in steel and rubber, and how we think of ourselves as people has changed since the practice of driving became widespread. Being able to physically, on your own, get in a car and drive away from a bad situation is different than telling a robotic vehicle to take you to a destination. Aside from the fact that you may not even *have* a destination, driving on your own, in the types of cars we've been driving throughout the twentieth and twenty-first centuries, means that where you go and how you get there is your own private business.

You could, if you wanted, throw this book and a suitcase in your car in the middle of the night and just take off, to wherever you want. That's a potent and exciting possibility to have just waiting there at the back of your mind, a last-ditch fail-safe. When the only way to move around on your own is passively, inside a machine that communicates your every destination and plan with a larger, interconnected network of other robotic cars, it's not the same. Your trips or plans will never really be private or spontaneous, and as a result we will lose some ability to be private and spontaneous.

The most obvious thing we'll be giving up in a switch to robotic vehicles is the physical pleasure you can get from driving.

Also, get your mind out of the gutter, you perv, you know what I mean, and if you don't, well, then, I guess I pity you, a little bit. Even though the majority of drivers have had the joys of driving beaten out of them by years and years of slow, tedious, traffic-choked commutes, I'm sure all drivers, at least at one point in their driving life, have felt the visceral thrill of taking a corner just right or enjoyed the pit of the gut excitement of really stomping the gas of a fast car.

There's no reason robotic vehicles can't be fast or handle with remarkable precision, but what does that mean if nobody is actually doing the driving? I know you're likely sick of this car-as-prosthetic thing, but there's really something to it; the way you can feel the weight shift in a car as you take a turn, the way you feel a car nose in a little as you brake, or settle back as you give it more throttle. These are all sensations communicated to you through the car via the body, via your ass on that seat, by the way you feel gravity tugging your person, all very basic, tangible sensations. Nobody drives by the instruments when you're driving for fun; it's like any athletic/physical activity we find fun. The actions themselves, the motion and physics of it, are things we respond to and enjoy. And while riding along as a passenger in a car on a track is fun, it's acres different than the sensations of driving yourself.

Driving a sports car properly set up on a real track is the best way to experience these sensations, but most people don't have the time or resources to avail themselves of track driving, with its expensive equipment and attitudes and demanding requirements. But what's great about the widespread and ubiquitous nature of cars and driving is that pretty much anyone in any car can find a time and place to enjoy the physical sensations of driving. Plus, there are a lot of low-barrier-to-entry amateur performance/

pleasure driving opportunities in America, from run-what-you-brung drag strip open nights and off-road courses to entering a car you're about to scrap in a local demolition derby because why the hell not.

Humans are pretty capable of taking things we have to do out of necessity and making highly pleasurable activities out of them. Look at eating, for example. Sure, we could just drink something like that Soylent goop and take vitamins so we don't get scurvy like some filthy pirate, but most of us want to *enjoy* the process of eating, and not just subsist.

Driving can be similar. Many or most people have to drive just to get to work and navigate through their lives, but it's also entirely possible to use driving skills and cars for nonpractical, pleasurable purposes.

CHAPTER 9

WILL THEY BE LIKE YOUR DOG?

THE ONE THING WE CAN SAY, AT THIS TIME, WITH ABSOLUTE certainty about how autonomous cars will change our lives, culture, and society is that, really, we have no fucking idea. That's why I'm extremely hesitant to make predictions. As a miserable old person, I recall a time when cellular phones were either imprisoned in cars or toted around in black pleather satchels. If you had told me that one day those phones would be combined with cameras, music players, and powerful computers that could interface with a global network and would be used for people to send immediate, up-to-the-moment pictures of their meals and every action taken by their cat, I most likely would have counseled you to seek help from your clergyperson.

But here we are.

Now, even though I think predictions are generally doomed to seem either ridiculous or incredibly lucky, that doesn't mean I'm not going to do pretty much exactly that right now, but let me hedge my bets and just say I'm speculating here. Because, really, I think there's a hell of a lot to speculate about regarding how our interactions with vehicles that can drive on their own will be. I mean, let's consider for a moment how incredibly rich, complex, and segmented the way humans deal with normal cars is. As I believe I've already suggested until you were sick of hearing it, we don't treat cars like we do the other machines in our lives. There are entire subcultures dedicated to various types and styles of cars and ways of interacting with cars. Cars evoke incredibly emotional responses in people, and they do all this while remaining inert lumps of metal, rubber, and plastic when nobody is actively controlling them. What will happen when we have cars that can move and react independently, with potentially quite complex behaviors?

I think we'd be foolish to think that won't mean something.

Even before we consider how we'll be interacting with these two-ton mobile computers, it's worth thinking about what sorts of additional capabilities these things could bring to our lives—things beyond the normal range of what we expect to be doing with autonomous vehicles, novel things that take advantage of the unique capabilities of this new breed of machine. They're not really cars, so why would we only use them the same way we've been using cars for the past century?

Fundamentally, an autonomous car is just a bunch of computers combined with a car. The computer that's driving the car (its brain) is relying on data streams from cameras, sensors, and GPS systems (the body parts), but in the end, it's all just numbers. And because of the nature of computers, this means there are opportunities to make content for autonomous cars, because autonomous cars can be playback devices. I'll explain.

It should be technically and theoretically possible for an autonomous car to drive with input not just from its array of cameras and sensors, but from a set of prerecorded data from another car. The data that can be recorded—and almost every drive-by-wire car made today is recorded—is significant: throttle position, steering wheel angle, brake pressure, weight balance, speed, wheel slippage, and, of course, GPS data about where the car goes. If you're wondering why the hell anyone would want to do this, think about these possibilities:

- You could go to a particular track and download the fastest lap ever recorded on that track, and "replay" that lap in your car. Famous drivers could record hot laps in cars, and you could download and replay their lap—every action they made, every jiggle of the wheel, every trail-braking (you know, a bit of brake pressure to settle the car's balance in hard driving) foot on the pedal, following the exact line they took through every corner, everything—in your own car. Well, modified to adapt to the performance limits of whatever you're driving, of course.

- Even if you weren't in the exact same kind of car, this would be an incredible way to learn a track, or just simply enjoy the thrill of whipping around the track with a pro driver at the "wheel."

- You could download entire, curated road trips. Let's say you wanted to take Charlie Day's Amazing Corn Dog Tour Of America, a road trip he took and recorded where he traversed thirty-seven states to find the best corn dogs in the US. You could download the trip, perhaps with some sort of audio commentary track or music playlist, and set off: the GPS path of the original trip would play, complete with stopping at selected corn dog palaces.

- You could tackle a challenging off-road course even if you've never done it before. Assuming you have a vehicle with the right equipment, you could download and play back an off-road drive from an expert who will not get you stuck, or at least will give you a fighting chance at not getting so stuck. Actually, this idea is already being sort of explored by Land Rover, as a way to help people drive off-road. It's not the same as downloading entire drives, but it's close in that it's providing a skill set to the car to augment and/or replace the human.

- You could go to a new city as a tourist and download a tour of the city, right to your car. There could be a huge number of very specific tours, planned and plotted by experts in a given field. Hell, I could have made a pretty good tour of all the interesting and weird cars on the Eastside of LA. There could be architecture tours, food tours, booze tours, whatever. And since the car is driving, you could gawk and drink as much as you'd like.

- Hypermilers, traffic avoiders, speed junkies, and all sorts of other specific-driving-interest groups could create and share or sell routes that cater to their particular interests. For example, if you have an electric car and really need to find a route that lets you squeeze the maximum range out of your batteries, you could find and download a recorded drive from an electric car hypermiling expert who has the patience and skills to make that possible. You may have a long, annoying commute and be able to join an online group of people with similar commutes who work together to find the optimal, least-trafficked, most scenic, or quickest routes, and they could exchange and share their recorded drive files.

DONUTS_IN_PARKINGLOT.DRV
64 KB

HotLap_JensenButton.drv
64 KB

LA_Car_Tour_3.drv
64 KB

MYBESTAUTOCROSS.drv
64 KB

Reykjavik_Tour_Bjork.drv
64 KB

RoadTrip_1.drv
64 KB

WalterMondaleEroticAmerica.drv
64 KB

wreckintotreeDONOTUSE.drv
64 KB

There are far more possibilities here that I haven't even dived into; stored autocross runs, forensic use by police to play back the drive of a criminal's car, recordings of famous movie drives, like the chase scene from *Bullitt* (you'd probably have to block off roads for that one), synchronized automotive-dance programs for many cars in big, empty parking lots, possibly for weddings, and I'm sure lots more.

Of course, these recorded playback drives would have to be subject to the entropy of the real world. Sometimes traffic and environmental conditions won't allow for the exact playback of a recorded drive, and in those cases the car's existing camera and sensor input would take precedence, and adapt the playback accordingly, to keep everyone safe. There's no reason that wouldn't be technically possible.

Ideally, I'd like one universal standard for automotive playback content, so we don't end up with a bunch of incompatible app store-like situations for every automaker. I'm thinking there'd be some sort of online repository—possibly many—of these, with provisions for selling for-pay downloadable driving files, as well as an open marketplace for user-generated driving files for people to share. There would need to be vetting processes, of course, so people can't develop malware-driving files that send the cars on spiraling death-rampages, but the cars themselves should also be able to evaluate for safety concerns any driving file that gets loaded, and to adapt the files to the safe limits of the particular car.

I'm sure a lot of people will comment that this will open cars up even more to hacking and security issues, and they'd probably be right. There should always be a quick, easy way to stop the playback of a driving file if, somehow, the car's own danger-detection systems miss something or are compromised. An easy emergency shutdown is probably a good idea to have in an autonomous car.

Okay, after just typing that, part of my insecure brain is wondering if maybe this is all a terrible idea. No, I'm sticking to my guns, here. This could be really great, and open up all kinds of possibilities for autonomous cars, and could be something that helps preserve the joy of driving even in an age when there's no driving to be done.

Also, it creates a new job perfect for people who love driving: Interesting Drive Experience Content Creator. That sounds like a pretty sweet gig. Your job would be to find interesting travel experiences, undertake the trips, and record them in your car. The interesting thing about this is the potential for nuance in the way the experience recorder chooses to drive.

The experience creator could make a voyage based on, say, the idea of the old Cannonball Run races; in that case, your car would be instructed to drive fast (within legal limits, I assume, but I also bet hacked versions of such drives, manipulated to break the cars' safety restrictions, would be available on whispered about websites) and the overall tone would be one of a cross-country rally. It could be exciting and fun in a sort of high-pressure way, and would be exactly what many other people wouldn't want at all. For those people, drives through, say, Southern California wine country could be made available, with the driver recording a pleasant leisurely pace, with the car slowing down at the exact right times and places to, say, view a sunset, or, with an even greater degree of coordination, meet a delegation from a given winery to tour the grounds or sample wine or whatever the hell people do when they go to wineries.

With the car always knowing its location and having the hardware needed to be in communication with the internet, these recorded drives could be remarkably interactive; other cars playing back related drive files could be contacted for rendezvous, or locations of interest could be alerted to your arrival, and preparations made (and likely paid for) ahead of time and accordingly. A whole trip could be orchestrated to a degree previously reserved for heads of state, popes, or, perhaps, Beyoncé.

It's also easy to imagine driving content would include the sorts of things that amateurs would create and share as well. Groups would spring up for people with common interests, as the internet has been creating since the pre-World Wide Web days of USENET groups or dialing into BBS systems. Consider the hypermiler example we talked about earlier; it's easy to imagine how informal competitions could start where drivers record their inputs to try and get the absolute most efficient driving method to extract the most possible miles of range from a battery or tank of fuel.

That is, if they have cars that also allow human input *and* they still know how to actually drive, two criteria I wouldn't necessarily count on in the long-term future.

Of course, the idea of a car as a playback device for travel, while interesting, doesn't begin to address the truly autonomous nature of robotic vehicles. If we assume that all of the current hurdles and bugs and issues regarding how autonomous vehicles can drive without running into things are solved, there's still the question of exactly how these cars will behave, and how we will react to that behavior.

If we assume that at some point in the future, robotic vehicles of all kinds will be common, we, as humans living in a modern society, will undoubtedly have a great deal of contact with these machines, in much the same way our counterparts in the 1800s would have had a great deal of horse contact. The robotic vehicles will use very complex algorithms to do what they do—in fact, if they're to prove effective at all, those complex algorithms will have to do some extremely complex things, that would likely fall under the category of what we currently call "artificial intelligence," though, to be fair, no matter how complex it gets, it's really not analogous to human (or even animal) intelligence as we understand it.

Even so, these systems can become complex and obtuse enough that our non-artificially intelligent brains will do what they always do, and superimpose a layer of anthropomorphism on everything these driving robots do. No matter how much we may rationally understand about these cars, we can't help it.

Consider this, for example: what if a company builds a robotic city car with a target market of crowded urban centers filled with pedestrians. This robotic car has programming specially tailored for this environment, and is known for being very good at pedestrian avoidance and behavioral predictions. A side effect of this is that at nearly every stop sign or traffic light, the car will furiously scan the pedestrians in its near environment and attempt to predict their behavior. As a result, these cars will tend to "quake" a lot when stopped, making constant micro adjustments and slight motions to gauge pedestrian response in order for its behavior-prediction algorithms to work.

If you were standing on a corner watching the car, it would appear to be scared shitless.

The car, of course, isn't scared at all—it's just moving its motors for very specific reasons, and the result is it looks like it's hesitant and shaking. This, of course, could be a marketing disaster for whatever company built this car, because who wants to be driven by a terrified car? Nobody, that's who.

Of course, this car doesn't exist (at least not yet), but if we consider it as a thought experiment it reveals something about human nature and the machines we build: we constantly, almost instinctively, impose emotions and other human traits on machines. That means automakers may decide that they need to figure out desired perceived behavioral profiles for the cars they develop. Based on what we've seen of auto design in the early twenty-first century, it would be easy to think that the prevailing trend would be to design cars that convey an air of aggression, a threatening, forceful demeanor. Maybe this would be accomplished via stance, or how it lunges away from traffic lights, or refuses to back up or yield when approached by another car (within predefined safety parameters, of course).

As tempting as this might be to auto designers, looking for something comfortable to fall back on, as we discussed in the chapter about the design of self-driving cars, chances are people won't be comfortable with a machine that no human controls that behaves in an aggressive manner. Well, at least not at first; as trust in these machines builds over time, it won't be surprising to find that there are people out there who would want to own a robotic car that acts like an asshole.

That also suggests that there could be a market for specific behavior types for robotic cars. The nature of humans is already such that we'll be anthropomorphizing—or whatever the equivalent word is that suggests we'll be perceiving these vehicles like we perceive animals (I just Googled it and I think the term is "zoomorphism")—these robotic vehicles all the time, and seeing "personalities" in very technical behaviors. That's just what people do, and have always done.

We already form strong bonds with our cars, and often anthropomorphize a normal, human-driven car's "moods," depending on how the car is running or how it reacts to our input, even though we rationally know the results of those reactions are entirely due to mechanics and physics, perhaps a failing part or a lapse in maintenance. We know that the way the car is reacting has nothing to do with emotion, but many of us still get a pang of gut-level sympathy when we feel an engine misfiring, and we feel that same held-breath tension as a reluctant car cranks and cranks trying to start, and a visceral sense of relief when it does.

We know we're going to treat cars that can drive on their own like they have distinct personalities, and you just know there will be people who insist their cars behave with what they perceive as emotions or have perceived traits far beyond what the vehicle's actual programming is capable of. I suspect that entire channels on YouTube (or whatever the futuristic equivalent may be) dedicated to robotic vehicles doing strange, unexpected, cute, dangerous, or otherwise funny things will be quite popular.

Still, whether programmed to act like assholes or not, robotic vehicles will absolutely be programmed in such a way that has safety as its first priority, and will take any steps needed to avoid harming people. While this absolutely makes sense from a safety standpoint, it also means that the inherent good nature of these machines could be exploited by people whose natures are considerably less good. The level of alarm we can play with here is really pretty broad. On the innocuous end, it's very easy to imagine how robotic vehicles could be pranked by taking advantage of their collision avoidance systems. A few years ago[65] I imagined a whole set of robotic-vehicle pranks that the troubled teens of the future could enjoy:

1. CAR HERDING!

This one, like most of these, uses the collision avoidance sensors and cameras almost every autonomous car will have. And while I can't be certain this (or any of these) will work, it might!

For this one, you'll need a group of friends, say six or so. You'll spot an innocent robot car hunting around for a parking place, and then surround it in a loose, wide, circle. Various members of the group will approach the car's hazard-avoidance sensors (usually at the corners and central front and rear) to force the car to stop and/or change direction to avoid you. (I was in a prototype Nissan in Japan that did just this.)

If everyone works together, you should be able to "herd" the car at low speed in any direction you want, as if herding a nervous livestock animal or something. I bet it'll be fun. I don't think this would ever be a popular way to steal a car because it seems really, really slow.

2. INSTANT UNWANTED AUTOCROSS!

This one takes just a bit more advanced planning, and a whole bunch of cones. You see a car heading your way, so you set up a quick autocross or slalom course with the cones, and force the car into it. You'll probably have to use the herding technique to get it into the coned-off area, but once it's in, if you set your cones right, the car's sensors will force it to navigate the autocross track as best it can.

Set one up in a parking lot and run multiple cars through, while taking wagers on times for some serious fun!

3. AUTO SUMO!

This one may be tricky, but should be worth it. You'll need a nice big empty parking lot or something similarly useful and two autonomous cars. Herd them close to each other, and then enclose them in a ring of cones. From there, try and get them to circle, chase, and go after each other by blocking/revealing the collision sensors. At some point, the cars should start reacting to each other. Each car should have a GPS destination set just beyond the other car, so they are forced to navigate past one another. The winner is the one who forces the opponent out of the ring first!

4. THE TRAP!

This one's also pretty straightforward. Track a car as it's driving, then surround it with objects (front first, so it stops) so that you've trapped it!

The thing about all of these admittedly silly ideas is that if such things are possible, so are actual crimes that take advantage of these same behavioral traits. It could be possible to force a robotic vehicle to stop, and with enough people or objects, box it in, and then rob the car (if unpassengered) or the people inside, or kidnap them, or any number of other terrible things.

Because the rules that govern the car's behavior prevent the car from putting anyone in harm, driving aggressively through the people surrounding the car likely wouldn't be an option, even though in a human-driven car, the driver may choose a path that could potentially harm others in the interest of self-preservation.

I'm by no means the only one to entertain these grim, dark thoughts about a robotic-vehicle future; a 2017 report from the United Kingdom's UCL Transport Institute titled *Social and Behavioral Questions Associated with Autonomous Vehicles*[66] explored many of these same ideas, and in one section even described a robot-vehicle trapping scenario in detail, in the form of an imagined *Daily Mail* article from 2026:

SCENE G: AUTOMATED MUGGING

Local Transport Today, Issue 13227, February 4–17 2026

Transport Select Committee announces inquiry on personal security and autonomous vehicles

The Transport Select Committee yesterday announced the launch of a new inquiry on personal security and autonomous vehicles, reflecting serious concerns raised by a wide range of motoring and consumer group organisations (see LTT passim). These relate to the vulnerability of occupants of fully autonomous vehicles following a series of high profile vehicle-jackings and personal muggings in wealthy, low-density areas at night—throwing into question the whole idea of "hands-free" driving.

AVs travelling down residential streets have been suddenly surrounded by groups of young men, wielding bars and bats. The vehicles come to a halt, to avoid causing injury, and then remain immobile while windows are smashed and occupants are threatened. Having suffered the fear and humiliation of the attack, the occupants are further angered by the vehicle's monitoring systems identifying damage and thereby refusing to restart so they can resume their journey.

The Mail on Sunday has added its weight to the campaign for action. As it reported last month, Sue Brown was returning home from a night out with friends and while her vehicle was passing a local park something was thrown in front of her car, which made an emergency stop. Immediately she was surrounded by four youths; one smashed a side window and demanded her necklace, watch and purse. "What could I do?" she recalls. "If I'd had my old manual car I'd have driven at them and they would have soon scattered! We bought this car as we were told it was a lot safer—my husband had an accident a few years ago and injured a cyclist—but this is just exchanging one type of danger for another. We traded it in the next day for a 'proper' car."

Now, before we get too panicky, it's worth considering all of the other mitigating factors that could be available here. First, robotic vehicles are going to be connected to one another and, very likely, the internet at all times. Video feed from the car's cameras could easily broadcast to law enforcement what's happening to the people inside either by them requesting it, or possibly via an automatic algorithm in the car itself that's specifically designed to recognize potentially hazardous situations and call for help.

In addition to calling for help, there's no reason to assume that robotic vehicles—especially expensive ones bought by rich people (you know, the high-value targets, with all their money and gadgets and probably tender, well-marbled flesh if we're talking post-apocalyptic cannibal gangs here)—will be defenseless. It's very likely that robotic vehicles will be equipped with earsplitting sirens or door handles that deliver Taser-level shocks or spinning chains on robotic arms or firearms or who the hell knows what other kind of intimidating and possibly sadistic weaponry.

Less movie-like, there could also be the option of an "ethical override" button of some kind that, when pressed, logs who pressed it, records the entire incident inside the car and outside, immediately alerts the police to what's happening and where, and allows a limited time during which the car can ignore ethical rules like not running anyone over to get the hell out of a dangerous situation—pretty much exactly what a human would do. I'm also imagining that when a car is placed into some sort of emergency ethical-override situation, it would flash angry red lights and announce warnings like WARNING VEHICLE NO LONGER ADHERES TO ETHICAL STANDARDS! WARNING! VEHICLE MAY KILL WITHOUT REMORSE!

Of course, once you have the option to override ethical rules, that introduces a whole host of other issues and brings up some other interesting questions: will robotic vehicles be effective in helping to prevent crime—or, depending on how you want to think about it, will they remove our freedom to decide to do illegal things, and, if so, will human-driven vehicles be kept around specifically because they're better for doing illicit things in?

Think about it; robotic vehicles will be in constant communication with other vehicles and likely cloud-based networks run by manufacturers, law enforcement agencies, and probably a whole host of advertising, market research, and commercial entities that have purchased the right to eavesdrop on your data from . . . who the hell knows. The upshot of all this is that in a robotic vehicle you are never alone, and where you choose to go is never truly private.

Will law enforcement agencies geo-fence (you know, mark off a given area via GPS data) certain parts of cities that are known for specific sorts of criminal activity? If you got in a robotic vehicle with the intent of, say, purchasing illegal drugs or seeking a prostitute or some similar sort of illegal activity, will police be made aware of it because of a specific destination you told your car to go to? It would be easy for a given city to do something like this; it's just data, after all, and which areas of a city are set to trigger certain reactions from authorities would be impossible for an outsider to know. Really, in a robotic vehicle, it's hard to imagine how you could do anything you wanted to keep private or secret, even if it's not illegal.

The very nature of a network-connected vehicle means that there's a data trail for everywhere you go. If you're having an affair or going to secret cult meetings or taking forbidden dance classes, there will be a way to find out.

Will people who really want privacy—whether they're planning to do illegal things or not—choose to continue to drive their own cars even as robotic vehicles become increasingly common? Will human-driven cars then take on an air of suspicion for a lot of people, steeped as they may become in the idea that they're ideal if you have something to hide?

Could you use an autonomous vehicle as a getaway car? I doubt it, at least not without some serious hacking. Which leads us to the big robotic elephant in the room, by far the most serious security issue we'll face with robotic vehicles, far more important and potentially dangerous than any band of sour-faced ne'er-do-wells that may try to surround your robotic vehicle and beat on it with stale baguettes or golf clubs: hacking.

If there's a computer connected to any sort of network, someone will figure out how to hack it and make it do things you don't want it to do. That's the nature of how things work. There's no magic security bullet. Security developers come up with ever more secure networks, more powerful encryption, new methods to keep computers safe and uncompromised, and hackers eventually find loopholes and cracks and ways into once seemingly secure machines. It's an arms race, and there's no reason to assume this back-and-forth will ever end. The only sensible course of action is to accept that hacking systems, critical systems, *can* happen, and try to prepare accordingly.

The problem with the eventual widespread adoption of robotic vehicles is that unlike computer crime today, when terrible things happen—like bank accounts get drained, private documents get leaked, or maybe the power goes out—when we have four-thou-sand-pound computers with motors and wheels around, hacking becomes much more dangerous, physically. A hacker who can compromise a given robotic vehicle could, as you can imagine, cause a hell of a lot of trouble if they gain access to even a subset of the car's systems involved in driving. Steering, throttle, or braking control, or even just the ability to impair or impede the usual flow of commands, could have devastating consequences.

Cars could also be hacked in less dramatic ways, like spreading ransomware-like viruses from car to car that hobble the vehicles' capabilities unless money is paid. Every shitty thing that hackers or malware has managed to do to personal computers could happen to robotic vehicles, because, fundamentally, they're just computers.

I think, in general, most security systems will be enough to avoid disaster, in much the same way that, for the most part, our networks of computers and phones and spy cams and internet-connected refrigerators generally work. But we'd be foolish to think that hacking of connected and autonomous vehicles won't happen. Because it will. Really, the only absolute fail-safe solutions we may be able to use are physical; big, obvious buttons on the interior and exterior of robotic vehicles for emergency power cut-off may be a last-ditch solution, but those also open up opportunities for abuse.

That's all really depressing. Let's think about something more fun: motorsport. I think even without drivers, there will be new forms of motorsport that come with the advent of robotic vehicles, because, if racing lawnmowers and Barbie Power Wheels has taught us anything, it's that people will race absolutely anything with wheels. Without human drivers, watching fully autonomous racing might be an experience more like watching greyhound racing, in that the "vehicle" that is racing has no jockey or driver, and the vehicle and the entity that controls the vehicle are one and the same, either a very skinny canid or a machine.

The human element of this sort of racing would be in the design of the machines racing, both physical design and algorithmic/programming design to control things like their driving style, behavior on the track, and ability to make effective decisions on the fly. To ensure that autonomous racing doesn't become boring or sterile, I think the key is to let the cars be as varied as possible.

This is sort of the opposite of Formula-type racing, where the specifications for the given cars are followed so closely that the abilities of the human racers remain of paramount importance. With that element gone, the path to allowing humans to really compete and excel is through design and engineering. With that in mind, a set of general parameters for the cars could be established such as: maximum power, torque, weight, and dimensions. Any solutions that a given team could come up with that fit the restrictions and general rules would be okay.

This would open the door to some really innovative designs and novel engineering solutions, as well as being a really exciting crucible for the development of real-time driving AI driving systems. A robotic race car with a "brain" that can accurately perceive the best time to take risks, for example, would prove highly effective on the track, and would have interesting applications in the real world as well. Racing fans would have favorite teams instead of drivers, and perhaps there would be standout superstar engineers, designers, or coders. Human racers might still prove valuable and involved in autonomous racing by helping to "train" the race car's algorithms. These racers could be living racers recording data from laps on a track, or the cars could be fed thousands and thousands of hours of races from the past, provided accurate data could be gathered from old racing footage, which, while I believe may be possible, sounds like a colossal pain.

Racing robotic vehicles also opens up a great number of possibilities because safety requirements would be dramatically less important than in human-driven racing. This could mean higher top speeds and more technical, dynamic—or just flat-out bonkers—tracks with ridiculous but exciting features like jumps, loops, and severe banked turns, or any number of obstacles—swirling blades or jets of flame. Really, there's no reason not to go absolutely batshit once the issue of driver safety is removed. This could also open up entirely new categories of motorsport that are hybrids of demolition derbies and ultrafast Formula 1 racing. You want to have cars racing at 300 mph while being able to launch missiles at competing cars? Sure, why not? You want hydraulic slingshots randomly flinging cars off the track? Mining truck-sized robotic monster trucks hunting down small fast cars? With enough money and a willing audience, anything could be possible.

In terms of more mainstream entertainment and culture, it's interesting to note that we've had many popular works involving autonomous vehicles for decades and decades. For whatever reason, the 1960s seemed to spawn a lot of the more famous examples of this genre, starting with Ian Fleming's book *Chitty Chitty Bang Bang* in 1964, the television series *My Mother the Car* in 1965, and, perhaps most famously, the 1968 Disney movie, *The Love Bug*, about a sentient Volkswagen named Herbie. In the 1980s, we had perhaps the next most famous self-driving, self-aware car, KITT, the Trans Am that helped launch David Hasselhoff's career, in the television show *Knight Rider*.

On the more sinister side there were the Stephen King horror books/movies about sentient vehicles, *Christine* and *Maximum Overdrive*. We've liked the idea of self-driving cars for a very long time, and these early examples give us some good insight into how we may incorporate them into our culture once they're real.

There's a number of different ways these cars were portrayed. Chitty Chitty Bang Bang, for example, was perhaps the most inscrutable. Chitty had very little actual communication with the people he interacted with, and primarily revealed remarkable new technical abilities, right at the moment they were most needed, like the power of flight or the ability to turn into a pretty effective boat. Chitty Chitty Bang Bang did all this without too much overt communication or emotion, though perhaps that can be attributed to his Edwardian era environment.

Herbie the Love Bug, a sentient 1963 Volkswagen Beetle

with quite a successful racing career, is another interesting example. While Herbie was only able to communicate in the most rudimentary manner—horn beeps, engine revving, opening/closing body parts, and outright actions—he managed to convey a pretty complex personality. Herbie demonstrated loyalty, pride, competitiveness, affection, playfulness, and even murderous rage; there's a scene in the movie where in a fit of jealous anger he smashes a Lamborghini 400 GT into a twisted wreck. (It's worth mentioning that in the movie the Lambo somehow becomes a smashed Jaguar E-Type, which they hoped we wouldn't notice.) Herbie even attempts suicide by trying to jump off the Golden Gate Bridge. By modern standards, it's sort of insane to think that all this was in a mainstream Disney kid's movie. If, say, Olaf from *Frozen* attempted to slit his wrists with an icicle, I feel like people would freak out, but back in 1968, having the cute little Beetle star of a movie attempt to take his own life was just fine.

Herbie's personality was powerful and distinct. He had an iron will, perhaps a nod to his Teutonic heritage, and would stop at nothing to achieve his goals, even if it meant considerable self-harm. He won the final race of the first movie only after he'd literally torn himself in half due to his extreme performance in the race. If there was any doubt, know this: you do not fuck with Herbie.

Herbie's personality and the way he's treated is an odd gray area between how we would treat a human and how we would treat an animal. He was never treated exactly the same way a human character would be, largely because of his lack of language, and it was implied that his understanding of things was perhaps a bit more simplistic than that of a human. Still, he was expected to and frequently demonstrated understanding of concepts and situations far beyond what an animal could be relied on to process effectively. Herbie did all this without the aid of CGI (I'm ignoring the much later *Herbie: Fully Loaded* atrocity of 2005) and with the most basic of actions. Culturally, we're extremely open and willing to accept the idea of a car that has agency and personality.

Knight Rider's KITT is, in some ways, a closer analogue to robotic cars in the near future, in the sense that KITT's sentience or artificial sentience simulation is clearly the result of very advanced technology. KITT is sentient because he's a highly advanced computer with the voice of that doctor from that show *St. Elsewhere*. KITT's sentience is due to artificial intelligence, the result of science and engineering. Herbie's is much more vague, and the sentience is never *really* explained. If this movie were made today, some inane backstory involving wizards or a special high-tech Awareness Chip or some other madness would be worked into the script. But the closest we get to an explanation comes from Buddy Hackett's character, Tennessee Steinmetz, a mechanic and sculptor:

> **Tennessee:** Well then, if everything you say about this car is true, it's already starting to happen.
>
> **Jim:** What's starting to happen?
>
> **Tennessee:** Us human beings. We had a chance to make something out of this world. We blew it. Okay. Another kind of civilization is gonna take a turn.
>
> **Jim:** Give me an 11-mil wrench.

Tennessee: I'm sitting up on top of this mountain, right? I'm surrounded by these gurus and swamis and monks, right?

Jim: Right.

Tennessee: I'm lookin' at my stomach. I'm knockin' back a little rice wine. Got some contemplation goin'; I see things like they are. I coulda told you all this was comin'.

Jim: What's coming?

Tennessee: Jim, it's happening right under our noses and we can't see it. We take machines and stuff 'em with information until they're smarter than we are. Take a car. Most guys spread more love and time and money on their car in a week than they do on their wife and kids in a year. Pretty soon, you know what? The machine starts to think it is somebody.[67]

That's as far as it ever gets explained, and as far as explanations go, it's a pretty thoughtful one. In fact, what Tennessee is describing is basically Kurzweil's singularity (in case you're not a big geek, this is the notion that a self-improving artificial "intelligent agent" will eventually surpass human intelligence), but much more fun.

Really, it didn't matter in Herbie's context, but for KITT, there was at least a television-grade-level of plausibility, one that's roughly accurate for what real robotic vehicles may be like, and, importantly, there was the addition of language.

Michael Knight talked to KITT. A lot, if I recall correctly. It's very likely that the user interfaces of many robotic cars will include voice communication, since cars already have that, and our phones do it, as well as standalone devices like Amazon's Alexa line of speakers/digital assistants/corporate eavesdropping gadgets. Additional levels of personality will likely be included in these voice-controlled interfaces, via choice of voices and the sorts of responses they're scripted with for replies and interactions, and these will likely go a long way in affecting how we view these machines in the wider context of our lives.

Entertainment has provided some templates for dealing with self-aware, self-driving cars, but robotic cars will also have a reciprocal effect on entertainment, literature, and other arts. You know that moment when you realized how many of your pre-cell-phone-era favorite movies' plots would have been over in ten minutes if the characters had cell phones? The same sort of thing is likely to happen once autonomous vehicles become commonplace. So many plot devices—the drunk driving crash, being stranded because you can't drive or don't know where to go, a character being indisposed or out of communication because they were driving, not having an extra character to go retrieve something or someone, or give something or tell something to someone remote, or any sort of logistical combination of people and transport needs—will likely be altered when vehicles can follow instructions and drive on their own.

Hopefully, this won't actually ruin all that many movies, and perhaps it could even make for some compelling remakes. I'm sure a remake of *Speed* in which Sandra Bullock could just *tell* the bus to stay at 50 mph would make for a much more low-key, relaxed character study movie, similar to *My Dinner with Andre*, which is what people really wanted, anyway.

CHAPTER 10

SAVE THE GEARHEADS

I DON'T SEE THE POINT IN HIDING ANYTHING FROM YOU, ESPE- cially since you've slogged through so much of this book, or at least exhibited the bold daring to skip all the way to the end to see who the murderer was. So, since I respect you so much and, yes, maybe have developed a little bit of a crush on you after all this time we've spent together (be cool about it, okay?), I'm going to tell you the truth: even though I've been thinking a lot about cars that drive themselves, I, a human, do not ever want to stop driving.

I'd say you can pry my steering wheel from my cold, dead hands, but the truth is I can't think of a worse way for either of us to spend an afternoon. Instead, let's think about this in a more positive way: I, like many, many other automobile enthusiasts, fetishists, gearheads, whatever, love the process of driving, and am unwilling to give it up, even if autonomous cars manage to become commonplace and foolproof.

The truth is that right now there's so much we don't know about the future of robotic vehicles. We don't exactly know when they'll be developed enough for mass deployment; we don't exactly know how much they'll cost or how many of them will be sold, or if they'll mostly be shared fleet vehicles or privately owned. We can guess, and there are plenty of studies and very confident people writing articles and books (you know, like this, but let's pretend worse, somehow) and talking on podcasts and television shows and whatever, but, really, we don't know. It hasn't happened yet.

I suppose one other thing we absolutely *do* know, though, is that however and whenever the autonomous revolution happens, it's going to happen while the roads are still covered with human-driven cars. With that in mind, we need to figure out how these two types of vehicles—human-driven and machine-driven—are going to play nice.

There's a lot of worry that maybe they won't play nice at all, and the mixing of human- and machine-driven vehicles will be a disaster.[68] A February 2017 study from the Governors Highway Safety Association puts the onus of responsibility on the human drivers, and suggests a plan of educating drivers, taking time and being cautious before passing laws, capturing as much data as possible from driving interactions (good and bad) between the two main categories of cars, and making sure to plan and coordinate with law enforcement agencies.[69] There's no way around this. There are far too many human-driven cars around to even pretend that they'll just go away once robotic vehicles become viable. We're going to be looking at a mix for quite a long time, decades at least, and I believe human driving should never entirely go away, anyway. This means we need to figure out how to make things work.

As far as who should accommodate whom when coexisting on the roads I think it makes sense to have a system guided by the principle that majority rules. In the early stages of deployment, as long as human-driven cars far outnumber robotic vehicles, it will be up to the robotic vehicles and their manufacturers and designers to have them work well on primarily human-driven roads. This means they should be programmed to expect a driving world that is likely to be somewhat chaotic. While vehicle-to-vehicle communications will be extremely important, they should be designed to operate well in situations where there are no other vehicles to communicate with. Robotic vehicles will, unfortunately for their developers, be coming into a world that is essentially full of potentially worst-case scenarios, and that's just how it is.

Eventually, yes, once there's a critical mass of robotic vehicles around, all communicating with one another and coordinating their actions and collective traffic to the best of their little algorithmic e-brains' abilities, life for these machines will get easier. But that's not how things will begin.

Also, in the early stages, it may prove helpful for robotic vehicles to make their presence very obvious. With this in mind, something like a federally mandated external robotic/autonomous vehicle warning lamp could prove effective.

It may sound a bit like I'm paranoid about a massive robocar uprising and I want to be able to see potential uprisers, but that's not it at all. I think being able to visually see where the autonomous cars are will help them be accepted into mainstream traffic, and provide tools to study their interactions with other autonomous cars, meat-driven cars, pedestrians, and all the other chaos of life on the roads. I also think nervous human drivers might feel more comfortable if there's an easy way to identify which cars are machine-driven, so, if they choose, they can safely avoid those cars. Even if it's not entirely rational for drivers to fear the machine-driven cars, there's no reason for drivers to feel paranoid, and a clear signal as to where the robotic cars are may prove to be a welcome gesture of openness and, to those conspiracy-minded drivers, hopefully a bit of evidence that there's not some motivation to keep them hidden as part of some grand, evil scheme, possibly run by the Freemasons.

Here's what I'm thinking: any car that has some level of autonomous capability—let's say Level 4 or better, where a human driver is no longer a requirement at all—would be required to have a roof-mounted light. This light should be visible from 360 degrees around the car, and, significantly, from above. I'm thinking the color of the light should be something not already in common use, which is trickier than you'd think.

Red, orange/amber, and clear/white are all taken by regular car indicators or driving and running lights, blue is for cops, and yellow is often construction equipment, which pretty much just leaves green and purple. Green is sometimes used by private security, and is, of course, the "go" color on traffic lights, so until someone figures out how to make a viable brown or gray light, I guess we'll have to go with purple. Purple is sometimes used by funeral cars, which is a pretty unfortunate association, but I still think it's the least commonly used lamp color.

AUTONOMY LAMP

Maybe the funeral cars and hearses can use a black light? Would that even work?

This lamp could also prove useful for proponents of robotic vehicles. For those who genuinely believe that autonomous cars are safe, and, even, their superiority over human drivers, those purple lights could be a beacon of perceived safety.

And, speaking of safety, drivers of emergency vehicles, like ambulances or fire engines, could use the lights to identify where autonomous cars are on a road, and, provided we come up with a standardized procedure for what autonomous cars should do when an emergency vehicle approaches, those drivers could use the autonomous cars as a predictable path to their destination. An ambulance driver approaching an autonomous car can count on a set series of actions for that car to take, something they can't guarantee when barreling at some rando in a rusty Cressida. This could help those vehicles get through traffic more quickly and safely.

The reason the light should be visible from above is for easy tracking of autonomous cars, and their patterns. I'm sure manufacturers and researchers can get GPS and other data from their cars to study how they're doing in various driving and traffic situations. That does not mean those manufacturers have to share that data with anyone else. The overhead-visible autonomy light would allow independent researchers to study the behaviors and habits of autonomous cars without being beholden to the manufacturers. Cameras mounted to buildings or drone-based cameras could monitor traffic in city streets and see which cars were robo-driven, and how they behaved. Very valuable data will likely be gathered by watching how the cars work in reality, and having a means to get this data independently of any on-car logging or anything like that could prove very useful. Police could use it in car chases. You could tell at a distance what sort of driver your Uber has, meat or metal, and decide if you want one or the other. With just a glance you'd know if a vintage car had been converted to autonomous control.

Also, depending on how the laws eventually shake out, it could prepare other drivers and/or law enforcement to know what to expect in a given car. It may become legal to have a totally unmanned car, or a car without a driver carrying someone's kids or dogs or elderly mother. A purple light will let people know there may not be a driving-age human adult in the vehicle.

Up until now, a car's driver could only be one thing: a human being (exceptions made for sassy orangutans). Now that there's another option, it's simply a matter of acknowledging, without judgment, that what's behind the wheel of a car is valuable information to convey.

Now, when the number of robotic vehicles on the roads reaches a certain critical mass, it's going to be up to us moist, clammy humans to be a bit more accommodating. I'm not really certain what this number would be, and I'm not really sure anyone is, at least not yet. If I had to guess, I'd say somewhere around 15 to 25 percent. I just made up that number, but it seems a reasonable guess. When there's enough robotic vehicles driving around that their vehicle-to-vehicle communications and optimizing are starting to make differences in traffic flow and overall driving behavior, it may be reasonable to require that human-driven cars get in on the conversation as well, by retrofitting human-driven cars—even potentially quite vintage ones—with vehicle-to-vehicle transceivers that would provide some crucial bits of information to the whole connected car landscape:

- Announce that there's a human-driven car and its location

- Provide data on speed, GPS location, throttle position, and if the brakes are actuated or not

- Turn signal actuation

- Maybe steering wheel position as well?

If human-driven cars broadcast such information on the same vehicle-to-vehicle network that all the connected robotic cars are using, those autonomous vehicles could be made aware of the fact that there are, say, six human-driven cars around me, so I should be aware that these vehicles cannot necessarily be counted on to behave like me or my robotic compatriots. They're crazy-ass humans, after all, the same species of being that spent time and money to produce the Police Academy series of movies.

With data like speed, throttle position, GPS coordinates, and steering input, the robotic vehicles could even estimate probable vectors of travel at any given moment for these human-driven cars, and, by reading if a given turn signal is activated, even get some limited sense of the planned immediate future actions of the human-driven vehicle.

Take for example a lane change on a crowded highway. If a robotic vehicle is in one lane and a human-driven car in the other, and the human-driven car wishes to change lanes, the robotic car would get a bit of warning if the human used their indicator, and, if the robotic car is in the human car's blind spot, could be prepared to slow down to allow the human car to get into the lane. Even if the human car didn't signal, but just swerved into the lane, the change in steering wheel angle could give the robotic car enough of a warning to prepare accordingly.

All of these signals could be gathered from almost any car, though it would take a bit of installation work for older cars without engine computers that provide this sort of information already. Almost every car made since the mid-2000s likely has all of these signals available via the CAN bus (the internal communications network used in modern cars), but even on cars like my old 1973 Beetle I can imagine how small sensors could be placed on the throttle linkage at the carburetor to read throttle position, the brake light switch could be used to indicate when the brakes are active, and a sensor on the steering shaft could give steering wheel position. It's certainly possible for almost any car.

This sort of thing will likely become even more important when (and if, I should say) the day comes that robotic vehicles compose the majority of road traffic. When (just read an implied "and if" for all these predictions, if you don't mind) this does happen, there will likely be a lot of other issues that will arise regarding human driving in a robotically driven world.

If we assume that a time will come when robotic vehicles outnumber human-driven vehicles, we can assume that the robotic vehicles will have proven themselves to be much safer than human-driven vehicles, as their proponents confidently predict will be the case every freaking time you get stuck with them on an elevator. We can also assume that the general public will have come to trust robotic vehicles, and, with that growing trust, I think we can also expect an erosion of trust in the abilities of humans to drive safely.

If the safety improvements are as dramatic as autonomous vehicle proponents predict, then there may very well be people who will feel that humans driving themselves is an absolutely insane risk to take, and there may be movements taken to outlaw it entirely.

Even if human-driven cars are not outlawed, their use and operation may be restricted in a number of ways. While we're still not exactly sure how insurance and responsibility for accidents will work for autonomous vehicles—is the carmaker at fault? The company that produced the software? The makers of the individual components? The owner/passenger?—we do know how insurance works for human-driven cars, and it's possible that in an autonomous world, insurance rates for driving your own car might be incredibly expensive, as human driving may come to be seen as a risky behavior.

It's possible that human-driven cars may be relegated to certain lanes, roads, or regions; it's possible that entire cities or states could decide to ban human-driven cars. This sort of future may be a grim one (from a gearhead's perspective) in that car ownership and driving could become like horse ownership and operation today—limited to certain areas of operation, like a limited and restricted set of roads and private tracks, expensive, and only for a relatively small elite. I hope this is not how things end up.

It's likely that enthusiasts of driving their own cars will have to band together and form lobbying agencies if a mostly autonomous future comes to pass. Such a group could, hypothetically, end up like the NRA: controversial because the group believes in and promotes the ownership and use of machines with the potential to kill people, even though I'm pretty sure there are a lot of other things you can use a car for.

While I wouldn't want the NRA's often eye-rolling TV channel, I would hope that people who love driving could have the clout and influence of the NRA. Something like a National Drivers' Association (NDA) might be a necessity if we want to ensure a future where people can drive their own cars on public roads and not be limited to private tracks.

Think how hard it could be to argue you're not at fault if you're a driver involved in a wreck with an autonomous car. It's easy to imagine that the assumption will always be against the human driver, and without some organized network of support and legal resources, we can imagine a future where the legal and financial risks associated with driving yourself make the prospect not worth it for many people who would otherwise love to keep driving.

There are some good arguments to be made in favor of keeping human-driven cars around, and keeping alive the necessary skills needed to drive cars on your own. The most obvious reason is security. Once most cars are autonomous and communicate with one another and require massive infrastructures—like the internet, a robust electrical grid, and GPS satellite constellations—then many, many new points of failure or targets for attack will be opened up. In the same way that the United States keeps a Strategic Petroleum Reserve to ensure that critical motor vehicles can be used even in a catastrophic event that deprives the United States of sources of fuel, a strategic reserve of human-driven vehicles—mostly emergency vehicles and military vehicles—should be kept around in case anything catastrophic happens to the infrastructures that robotic vehicles will need to operate.

If there's a virus that spreads among robotic vehicles that has the potential to cripple even a percentage of them, that's a huge deal, and would have a massive effect on the life and economy of America. There will always be a need for human-driven backups for crucial vehicular-based systems, and I don't see any reasonable way around that.

There's also the fact that no matter what happens with economies of scale or technological innovations, autonomous vehicles will require increasingly more complex hardware to work than human-driven cars, and as such will be more expensive.

In areas where car sharing or public transportation options are robust this may not prove to be much of an issue, but in more remote or rural areas there may still be demand for cheap, reliable human-driven cars. Perhaps at first the used car market will fill these needs, but eventually, there will possibly still be demand for new human-driven cars, if not in the United States, perhaps in countries with lower standards of living.

There are also circumstances and situations where an autonomous vehicle wouldn't make sense; certain utility vehicles and trucks, for example, that don't operate on set routes but are used more as mules or workhorses for moving, towing, and hauling things around worksites or factories, are probably still best served by having a human at the controls.

Vehicles designed for extreme off-roading or other very experiential motorsports or hobbies will also likely continue to be human-driven, since the act of *doing* those sorts of things is what makes the experiences worth having. The same goes for nearly any motorsport, really, like drag racing, track driving, drifting, or autocrossing. These use cars that are already not road legal, and almost exclusively as sporting equipment.

While participation in these sports may decline as the baseline level of driver's education declines, there's no reason to think these motorsports will go away just because normal commuter driving goes away, in the same way that we still have people who learn to race horses over a century after horses were a relevant method of transportation.

In this speculated future, it's worth wondering what will become of driver's education. When I was growing up, driver's ed was provided, for free, by the public school system. Of course, my school system also let seventeen-year-olds drive school buses, which is why there were so many beheaded mailboxes in our neighborhoods and why so many buses ended up on their sides. While letting someone with barely a year of experience drive your kids to and from school seems insane by modern standards of not-letting-your-kid-die, it shows just how expected and common it was for everyone to learn to drive. The percentage of teenagers getting their driver's licenses has fallen in recent years, with about 85 percent getting their license by age eighteen in 1996, falling to

73 percent in 2010, and about 67 percent in 2018.[70]

So, fewer people are learning to drive, for any number of reasons beyond the scope of this book, or—if I'm honest—my interest, because what the hell is the matter with a teen who doesn't want to drive? Wise up, dummies.

As increasing levels of car autonomy become more and more common, we'll also see an erosion in the actual skill level required to learn to drive. Driving will get progressively easier, culminating with the widespread adoption of Level 5 autonomy, when the need to learn how to drive effectively disappears. In some ways, the decreasing skill requirements of driving parallels that of personal computers, which went through similar steps in a far more compressed timeline: the first personal computers, like the Altair, required users to be able to build the machines or at least understand them down to their bare metal: assembly language, light-and-switch interfaces, and all that. This is like cars before the Model T. Soon after that, the Apple II brought plastic cases, keyboards, and TV displays to computers, but you still really had to know how to program (at least in BASIC) to do much. These were like cars that you still had to shift and know a bit about how they worked. Later in the '80s, purchased software became more common and the Apple II, Commodore 64, Atari, and other computers were able to be used productively with only a basic understanding of command-line operating systems. This would be similar to the start of the automatic transmission era. The Macintosh's graphical user interface and mouse made things even easier—this would be like today's highly assisted-driving cars—and today computers are ubiquitous, and we use them for everything, and no one really has to know how they work at all, in any way. That's the autonomous era.

Still, if we keep this analogy going, it's worth noting that people still learn to program computers—sometimes as a career, and sometimes just because they're interested and want to do it on their own, for a hobby, side projects, jobs, or even artistic reasons. In the same way, I think there will always be a critical mass of people who are interested enough in cars to want to learn how to drive.

There is a potential upside for human-driving in a robot-driven world. Such a world would mean the death of the boring car—at least the boring human-driven car. Well, maybe there's still a place for cheap human-driven cars, and they may be boring, but, as I mentioned before, those likely will be limited to the developing world or specific low-cost niches. Think about it, if you're going to build a human-driven car when there are autonomous options, what would be the point of a boring-to-drive human-driven car? There would be no point. Nobody is going to build human-driven Camrys or Priuses or similar, not very interesting cars in such a future. The companies that will survive to build human-driven cars will be companies that make cars that offer unique and engaging driving experiences. This shift to autonomy could prove to be the thing that keeps companies like Lotus, Morgan, Porsche, Lamborghini, McLaren, and other similar small, niche sports cars makers.

If you're going to bother to drive yourself in an autonomous-dominated future, you're going to drive something that you love driving. That thing you love to drive may very well be a vintage car, and I think these will always have a place in our society, as they have value to our culture as a whole, as artifacts of an extremely influential era in modern human civilization. Even outside museums, there will always be people interested in human-driven cars as collectors or history buffs; that's a culture that already exists and thrives, and will continue to do so. Vintage car collecting and culture spans the socioeconomic spectrum, and ranges from individuals restoring an old Beetle in a garage to people meticulously restoring one-of-a-kind Voisins to Concours quality so they can be shown at Pebble Beach.

I would expect enough interest in vintage cars to last well into an autonomous era for there to be "robotic chauffeurs" that could be retrofitted into human-driven cars—even extremely old ones—so they could be driven autonomously. If autonomous vehicles become truly widespread, it's possible that knowing how to drive might be uncommon enough that even people with vintage car collections may not know how to drive, or it may be difficult for museums or other institutions to find qualified drivers, so they'd need machines like these.

There are, of course, significant challenges and restrictions associated with this. The unit would need to be roughly humanoid in form, to occupy the driving position of the car. Since you can't always count on drive-by-wire controls in a car, especially a vintage one, physical interfaces with the car's controls would be needed.

For the pedals, the problem has already been solved by the auto towing and trailering submarket. There are products known as supplemental braking systems that are essentially robotic feet that press a towed vehicle's brake pedal. This same basic concept would be used by our robotic chauffeur, but instead of just the brake, a unit—connected via Bluetooth or another wireless link to the main torso unit—would actuate the brake, accelerator, and, if present, the clutch. The pedal unit, since it could be wirelessly controlled, could be repositioned on the vehicle's floor wherever needed. The steering wheel controls on the main body of the robot would need to be adjustable and secured to the wheel via some manner of clamp system. An extra arm on the robot could be trained to find the locations of other controls, like lights, wipers, and indicators.

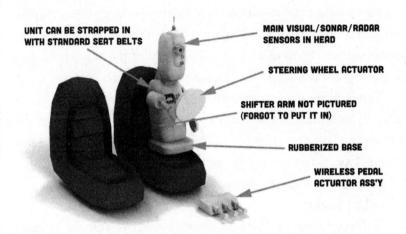

UNIT CAN BE STRAPPED IN
WITH STANDARD SEAT BELTS

MAIN VISUAL/SONAR/RADAR
SENSORS IN HEAD

STEERING WHEEL ACTUATOR

SHIFTER ARM NOT PICTURED
(FORGOT TO PUT IT IN)

RUBBERIZED BASE

WIRELESS PEDAL
ACTUATOR ASS'Y

The robot's main body would be placed in the driver's seat and secured with the original seat belts, if available. If not, supplemental restraints could be used to secure the unit in place. The bottom of the main unit would have a rubberized pad to eliminate sliding on even the most American of huge vinyl bench seats. If an OBD-II port (the connector through which modern cars can give access to their internal electronic nervous systems) is available, the robot could connect to it to get vehicle data, like speed and RPMs, to assist with driving. If it's an older vehicle, speed would be calculated via the robot's GPS system, and RPMs could be assessed either via audio cues or visual inspection of a tachometer, or some combination of both.

If the car requires shifting, there would likely need to be a third arm dedicated to controlling the gearshift. Initial setup would allow the robot to choose among standard patterns (like 4-speed H-pattern, 3-on-the-tree, or 6-speed) and perhaps a brief training session would be needed with a driver shifting to train the arm to find the gear's locations. Clutch point and pressure would likely require a similar training session to get right as well.

The head of the robot would contain the main visual, sonar, radar, and other sensors, though it's likely that additional sensors would need to be mounted on the exterior of the car, where they could communicate wirelessly with the main unit.

These autonomous driving units would have to be a bit more flexible and robust than units built into a single car, for the obvious reason that their inputs would have much more variable results than in cars whose characteristics are already known. Still, I think the technology is rapidly approaching a point where this could be possible.

This era of widespread human driving seems truly remarkable the more you think about it. In order to expand our horizons and conquer the impediments of physical space that potentially keep us from the resources we need, the opportunities we desire, or the people we love, we have developed machines—loud, smelly, fast, beautiful, ridiculous, practical machines—and trained ourselves to operate them.

Operating these machines is far beyond anything that individual, everyday humans have expected of themselves before. If you tried to explain to a king in the 1700s that one day a person from even some of the lowest rungs of society would be able to control a machine that flung them around the countryside at speeds of a mile a minute, that king probably would have had you beheaded for your ridiculous ideas and horrible impudence. But it happened.

Driving cars made us all into titans, gave us strong, fast bodies we could use to go places fast and carry huge loads wherever we needed. What the printing press did for ideas and brains, the car did for transportation and our bodies.

Cars won't go away once they become machines that can drive themselves, but they will definitely evolve into something new. No matter what happens, no matter how these new evolved robots may end up reshaping our world, there are those of us who will never want to let go of the experience of driving. There will always be people who can understand the benefits of the new era of robotic vehicles, who can appreciate the significant benefits in safety, the environmental improvements from shared-used autonomous cars and freeing our cities from the tyranny of parking decks and lots. People who can appreciate the convenience and accessibility and all the considerable advancements such machines have to offer, but who also feel, deep in their guts, the visceral pleasure of driving, the primal joy of acceleration, of the weight shift in a curving road, the satisfying *snik* of a perfect shift—people who accept the beautiful irrationality of human-driven cars, their excesses and failings, their absurdity and beauty, and will never, ever want to stop driving.

Now scoot, robot, and hand over the keys.

NOTES

1   Trevithick, Francis, *Life of Richard Trevithick: With an Account of His Inventions, Volume 1* (Cornwall: E. & F.N. Spon, 1872), p. 143, https://archive.org/details/liferichardtrev00trevgoog/page/n172.

2   Fuller, John, "Did da Vinci really sketch a primitive version of a car?" HowStuffWorks.com, https://auto.howstuffworks.com/da-vinci-car1.htm.

3   Milford, Frederick J., "US Navy Torpedoes," *Submarine Review*, April 1996, http://www.webcitation.org/5kmz4U6ZZ.

4   Kirby, Geoff, "Navies in Transition," *Journal of the Royal Navy Scientific Service*, Vol 27, No 1, https://archive.is/DIiw.

5   Felton, Ryan, "The Man Who Tested the First Driverless Car in 1925 Had a Bizarre Feud with Harry Houdini," Jalopnik, https://jalopnik.com/the-man-who-tested-the-first-driverless-car-in -1925-had-1792312207.

6   *New York Times*, "Radio-Driven Auto Runs Down Escort," July 28, 1925, https://timesmachine.nytimes.com/timesmachine/1925/07/28/ 99984359.html?pageNumber=28.

7   Google Patents, Sperry, E. A., "Gyroscopic Apparatus," Patented June 13, 1916, https://patents.google.com/patent/US1186856.

8   Exo Cruiser, "'Mechanical Mike' (The Evolution of the Modern Airplane Autopilot)," https://dodlithr.blogspot.com/2013/07/ mechanical-mike-1930s-forefather-of.html.

9   Lehman, Staci, "The Pigeon-Guided Missiles and Bat Bombs of World War II," Gizmodo, https://gizmodo.com/the-pigeon-guided-missiles -and-bat-bombs-of-world-war-i-1477007090.

10 Moore, James and Nero, Paul, "Pigeon-Guided Missiles," Military History Matters, November 10, 2010, https://www.military-history.org/articles/pigeon-guided-missiles.htm.

11 Donnelly, Jim, "Ralph R. Teetor," *Hemmings Classic Car*, July 2009, https://www.hemmings.com/magazine/hcc/2009/07/Ralph-R--Teetor/1846418.html.

12 Sears, David, "The Sightless Visionary Who Invented Cruise Control," Smithsonian.com, March 8, 2018, https://www.smithsonianmag.com/innovation/sightless-visionary-who-invented-cruise-control-180968418/.

13 Old Car Brochures, "GM Corporate and Concepts, 1956 GM Firebird II," http://www.oldcarbrochures.com/static/NA/GM%20Corporate%20and%20Concepts/1956_GM_Firebird_II/1956%20Firebird%20II-12-13.html.

14 Hicks, Nancy, "Nebraska tested driverless car technology 60 years ago," *Lincoln Journal Star*, September 12, 2017, http://journalstar.com/news/local/govt-and-politics/nebraska-tested-driverless-car-technology-years-ago/article_a702fab9-cac3-5a6e-a95c-9b597fdab078.html.

15 Reynolds, John, "Cruising into the future," *Telegraph*, May 26, 2001, https://www.telegraph.co.uk/motoring/4750544/Cruising-into-the-future.html.

16 General Motors, *Key to the Future*, https://www.youtube.com/watch?v=Rx6keHpeYak.

17 Hoggett, Reuben, "Cybernetic Animals - Stanford Cart," cyberneticzoo.com, http://cyberneticzoo.com/cyberneticanimals/1960-stanford-cart-american/.

18 Moravec, Hans, "Obstacle Avoidance and Navigation in the Real World by a Seeing Robot Rover," Carnegie-Mellon University, Robotics Institute, September 2, 1980, http://www.frc.ri.cmu.edu/users/hpm/project.archive/robot.papers/1975.cart/1980.html.thesis/p02.html.

19 ITS Laboratory, Biography of Sadayuki Tsugawa, Dr. Eng., http://www-ie.meijo-u.ac.jp/~tsugawa/sub1.html.

20 Schaub, Alexander, *Robust Perception from Optical Sensors for Reactive Behaviors in Autonomous Robotic Vehicles*, Springer Verlag, Berlin, 2018, pp. 17–18.

21 Dickmanns, Ernst D., Dynamic Machine Vision, http://www.dynavision.de/.

22 "DARPA Announces Third Grand Challenge," May 1, 2006, https://www.grandchallenge.org/grandchallenge/docs/PR_UC_Announce_Update_12_06.pdf.

23 Tartan Racing, http://www.tartanracing.org/.

24 Hikita, Munenori, "An introduction to ultrasonic sensors for vehicle parking," newselectronics, May 12, 2010, http://www.newelectronics.co.uk/electronics-technology/an-introduction-to-ultrasonic-sensors-for-vehicle-parking/24966/.

25 Rudolph, Gert and Voelzke, Uwe, "Three Sensor Types Drive Autonomous Vehicles," *Sensors Online*, November 10, 2017, https://www.sensorsmag.com/components/three-sensor-types-drive-autonomous-vehicles.

26 Utah, J, "Driving Downtown Object Detection - Rodeo Drive - Los Angeles, USA," YouTube, July 1, 2017, https://www.youtube.com/watch?v=PgnsapPGaaw.

27 Wikipedia, "Edge Detection," https://en.wikipedia.org/wiki/Edge_detection.

28 Torchinsky, Jason, "Why Nissan Built Realistic Inflatable Versions of Its Most Popular Cars," Jalopnik, October 18, 2012, https://jalopnik.com/why-nissan-built-realistic-inflatable-versions-of-its-m-5952415.

29 Condliffe, Jamie, "This Image Is Why Self-Driving Cars Come Loaded with Many Types of Sensors," *MIT Technology Review*, July 21, 2017, https://www. technologyreview.com/s/608321/this-image-is-why-self-driving-cars-come-loaded-with-many-types-of-sensors/.

30 Antunes, João, "Performance over Price: Lumina's Novel Lidar Tech for Autonomous Vehicles," SPAR 3D, May 5, 2017, https://www.spar3d.com/news/lidar/performance-price-luminars-novel-lidar-tech-autonomous-vehicles/.

31 Dwivedi, Priya, "Tracking a self-driving car with high precision," Towards Data Science, April 30, 2017, https://towardsdatascience. com/helping-a-self-driving-car-localize-itself-88705f419e4a.

32 Kichun Jo; Yongwoo Jo; Jae Kyu Suhr; Ho Gi Jung; Myoungho Sunwoo, "Precise Localization of an Autonomous Car Based on Probabilistic Noise Models of Road Surface Marker Features Using Multiple Cameras," IEEE Transactions on Intelligent Transportaion Systems, vol, 16, 6, December 2015, https://ieeexplore.ieee.org/ document/7160754/.

33 Silver, David, "How Self-Driving Cars Work," Medium, December 14, 2017, https://medium.com/udacity/how-self-driving-cars-work -f77c49dca47e.

34 Website of the Australian Government Department of Infrastructure, Regional Development and Cities, https://infrastructure.gov.au/ vehicles/mv_standards_act/files/Sub136_Austroads.pdf.

35 "Full Self-Driving Hardware on All Cars," Tesla.com, https://www.tesla.com/autopilot.

36 Thatcham Research Press Release, "Automated Driving Hype is Dangerously Confusing Drivers," October 18, 2018, https://news. thatcham.org/pressreleases/autonomous-driving-hype-is-dangerously -confusing-drivers-study-reveals-2767283.

37 Felton, Ryan, "Tesla Says Autopilot Was On Before Fatal Model X Crash, But That Driver Didn't Abide Warnings," Jalopnik, March 30, 2018, https://jalopnik.com/tesla-admits-autopilot-was-on-before-fatal -model-x-cras-1824224176.

38 Noyes, Dan, "Victim Who Dies in Tesla Crash Had Complained About Autopilot," ABC7 News, March 28, 2018, http://abc7news.com/ automotive/i-team-exclusive-victim-who-died-in-tesla-crash-had -complained-about-auto-pilot/3275600/.

39 Westbrook, Justin, "Tesla Blames Driver in Fatal Model X Autopilot Crash As Family Considers Legal Action," Jalopnik, April 11, 2018, https://jalopnik.com/tesla-blames-driver-in-fatal-model-x-autopilot -crash-as-1825193432.

40   "Discover Cadillac," Super Cruise, https://www.cadillac.com/world
     -of-cadillac/innovation/super-cruise.

41   King, Alanis, "Stop Doing This Shit with Autonomous Cars,"
     Jalopnik, January 15, 2018, https://jalopnik.com/stop-doing-this-shit
     -with-semi-autonomous-cars-1822090627.

42   Davies, Alex, "Ford's Working on a Remote Control for Your Car,"
     Wired, January 26, 2015, https://www.wired.com/2015/01/
     fords-working-remote-control-car/.

43   Coxworth, Ben, "Full Size Remote-Control Cars - Coming Soon to a
     Road Near You?" *New Atlas,* July 30, 2013, http://www.gizmag.com/
     remote-control-cars/28521/.

44   Hawkins, Andrew J., "Elon Musk Still Doesn't Think Lidar is
     Necessary for Fully Driverless Cars," The Verge, February 7, 2018,
     https://www.theverge.com/2018/2/7/16988628/elon-musk-lidar-self
     -driving-car-tesla.

45   Orlove, Raphael, "Angry Owners Sue Tesla for Using Them as Beta
     Testers of 'Dangerously Defective' Autopilot," Jalopnik, April 20,
     2017, https://jalopnik.com/angry-owners-sue-tesla-for-using-them
     -as-beta-testers-o-1794503348.

46   "Do We Need Asimov's Laws?" *MIT Technology Review,* May 16,
     2014, https://www.technologyreview.com/s/527336/do-we-need
     -asimovs-laws/.

47   Asimov, Isaac, "Sally," *Fantastic, Vol 2, No. 3,* May–June, 1953,
     pp. 34–50, https://archive.org/stream/Fantastic_v02n03_1953-05-
     06#page/n33/mode/2up.

48   Asimov, *Sally,* pp. 37–38.

49   Higgins, Tim, "The End of Car Ownership," *Wall Street Journal,*
     June 20, 2017, https://www.wsj.com/articles/the-end-of-car-ownership
     -1498011001.

50   T. S., "Why Driverless Cars Will Be Mostly Shared, Not Owned,"
     Economist, March 5, 2018, https://www.economist.com/blogs/
     economist-explains/2018/03/economist-explains-1.

51 Garfield, Leanna, "Only 20% of Americans Will Own a Car in 15 Years, Study Finds," *Business Insider*, May 4, 2017, http://www. businessinsider.com/no-one-will-own-a-car-in-the-future-2017-5.

52 Torchinsky, Jason, "I Have a Radical But Possible Idea What the Apple Car Will Be," *Jalopnik*, July 21, 2015, https://jalopnik.com/i-have-a-radical-but-possible-idea -what-the-apple-car-w-1719352935.

53 D'Olimpio, Laura, "The Trolley Dilemma: Would You Kill One Person to Save Five?" The Conversation, June 2, 2016, https://theconversation.com/the-trolley-dilemma-would-you-kill-one -person-to-save-five-57111.

54 Thompson, Judith, "Killing, Letting Die, and the Trolley Problem," *The Monist: An International Journal of General Philosophical Inquiry*, vol. 59, 1976, https://learning.hccs.edu/faculty/david.poston/ phil1301.80361/readings-for-march-31/JJ%20Thomson%20-%20 Killing-%20Letting%20Die-%20and%20the%20Trolley%20 Problem.pdf.

55 Khazan, Olga, "Is One of the Most Popular Psychology Experiments Worthless?" *The Atlantic*, July 24, 2014.

56 Thompson, "Killing, Letting Die, and the Trolley Problem," pp. 207–208.

57 Federal Ministry of Transport and Digital Infrastructure, German Government Ethics Commission on Automated and Connected Driving, Report, June 2017, pp. 10–13, http://www.bmvi.de/ SharedDocs/EN/publications/report-ethics-commission. pdf?__blob=publicationFile.

58 Torchinsky, Jason, "Should Autonomous Cars Be Forced to Save Lives in an Emergency?" *Jalopnik*, April 11, 2017, https://jalopnik.com/ should-autonomous-cars-be-forced-to-save-lives-in-an-em-1794121546.

59 Straight Dope Science Advisory Board, "Can Cops Really Commandeer Cars?" The Straight Dope, April 25, 2006, http://www. straightdope.com/columns/read/2247/can-cops-really-commandeer-cars.

60 Siler, Wes, "Science Shows People Prefer Angry, Aggressive Cars,"
    Jalopnik, October 7, 2008, https://jalopnik.com/5060127/
    science-shows-people-prefer-angry-aggressive-cars.

61 Warren, Tamara, "Google's Self-Driving Car Design Boss Speaks on
    Her Strategy," The Verge, October 25, 2016, https://www.theverge.
    com/2016/10/25/13307364/google-self-driving-car-design-yoojung
    -ahn-interview.

62 Torchinsky, Jason, "Honda Once Made a Car Specifically For People
    to Bone In," Jalopnik, November 2, 2015, https://jalopnik.com/
    honda-once-made-a-car-specifically-for-people-to-bone-i-1740050375.

63 Hanlon, Mike, "Honda's Fuya-jo Party on Wheels Concept,"
    New Atlas, April 16, 2005, https://newatlas.com/go/3950/.

64 "Branding Nano as the Cheapest Car Was a Big Mistake, Says
    Ratan Tata," India Today Online, November 30, 2013,
    https://www.indiatoday.in/business/story/nano-branding-big-mistake
    -ratan-tata-219211-2013-11-30.

65 Torchinsky, Jason, "Here's How to Prank Autonomous Cars When
    They Come," Jalopnik, July 25, 2013, https://jalopnik.com/
    heres-how-to-prank-autonomous-cars-when-they-come-874123410.

66 Cohen, Tom; Jones, Peter; and Cavoli, Clemence, Social and
    Behavioural Questions Associated with Autonomous Vehicles,
    Scoping Study by UCL Transport Institute, Final Report, London,
    Department for Transport, January 2017, https://assets.publishing.
    service.gov.uk/government/uploads/system/uploads/attachment_data/
    file/585545/social-and-behavioural-questions-associated-with-
    automated-vehicles-final-report.pdf.

67 Walsh, Bill. The Love Bug screenplay, 1968, Disney.

68 Felton, Ryan, "A Mix of Human Driven and Robot Cars on the
    Road Will Probably Be a Disaster," Jalopnik February 3, 2017,
    https://jalopnik.com/a-mix-of-human-driven-and-robot-cars-on-the
    -road-will-p-1791978251.

69 Governor's Highway Safety Association, "Driver Behavior Paramount as Autonomous Vehicles Introduced," New Release, February 2, 2017, https://www.ghsa.org/resources/av17-release.

70 Valeii, Kathi, "Kids Aren't Taking Driver's Ed Anymore," *The Week*, January 22, 2018, http://theweek.com/articles/749609/ kids-arent-taking-drivers-ed-anymore.